Frozen Section Library

Series Editor
Philip T. Cagle, MD
Houston, Texas, USA

For further volumes, go to
http://www.springer.com/series/7869

Frozen Section Library: Lymph Nodes

Edited by

Cherie H. Dunphy

Full Professor & Executive Director of Hematopathology
Department of Pathology & Laboratory Medicine
University of North Carolina Hospitals
Chapel Hill, NC, USA

Springer

Cherie H. Dunphy, MD
Professor & Executive Director
of Hematopathology and Hematopathology
Fellowship University of North Carolina
Department of Pathology & Laboratory Medicine
Chapel Hill, NC, USA
cdunphy@unch.unc.edu

ISSN 1868-4157 e-ISSN 1868-4165
ISBN 978-1-4614-1252-6 e-ISBN 978-1-4614-1253-3
DOI 10.1007/978-1-4614-1253-3
Springer New York Dordrecht Heidelberg London

Library of Congress Control Number: 2011939747

Printed on acid-free paper

Springer is part of Springer Science+Business Media (www.springer.com)

Preface

The Frozen Section Library series is to provide convenient, user-friendly handbooks for each organ system to expedite use in the hurried frozen section situation. This monograph, *Frozen Section Library: Lymph Node,* is a volume in the Frozen Section Library series. The Lymph Node volume discusses intraoperative approaches to evaluating both lymph node and extranodal tissues for hematolymphoid disorders. Since frozen sections are not routinely recommended for evaluating tissues for hematolymphoid disorders, this volume also highlights the usefulness of other techniques, such as touch preparation cytology in the frozen section room. In addition, other techniques, such as fine needle aspiration (FNA) cytology, as well as FNA biopsy are discussed as alternative approaches. This volume also suggests proper handling for subsequent ancillary studies, including flow cytometric, cytogenetic, and molecular studies. In addition, this volume discusses evaluation of lymph nodes for non-hematolymphoid malignancies and benign mimics.

This book serves as a very useful resource for physicians in the frozen section room and in intraoperative consultation situations dealing with, and interested in, this very complex field of diagnostic pathology. This user-friendly handbook focuses on practical diagnostic issues and is divided into chapters that emphasize the common questions a pathologist must answer during frozen section examination and provides guidance for the differential diagnosis of various histologic patterns. Special emphasis is given to the limitations of frozen section diagnosis in lymph node pathology. The chapters are written by experts in their fields and include the most up to date scientific information. The book targets on pathologists, residents, and fellows, who diagnose pathologic processes involving lymph nodes as well as hematolymphoid disorders involving extanodal tissues.

Although this volume addresses technical and interpretive issues involved in intraoperative consultations, one must remember that clinical context frames differential diagnoses, and good clinical histories help the pathologist direct diagnostic attention appropriately. When key information is lacking, differential diagnoses become unnecessarily narrow or broad, and interpretation suffers. It may be unreasonable to expect the pathologist to accurately classify an abnormal cell population in the absence of appropriate clinical history. Particularly important is any prior history of malignancy, as the possibility of a new metastasis from a historically remote neoplasm must be considered in the differential diagnosis of a new metastatic process. Although the impetus for clarity, completeness, and accuracy of diagnosis ultimately lies with the pathologist, the clinical team must provide an accurate account of relevant history and an intraoperative diagnosis should not be rendered in a clinical vacuum. Because one cannot always rely on this level of communication, a frozen section room setup with ready access to an online medical record can help avoid some of the shortcomings of histories provided on pathology requisitions.

Chapel Hill, NC, USA Cherie H. Dunphy, MD

Contents

Contributors

Cherie H. Dunphy, MD
Department of Pathology and Laboratory Medicine,
UNC School of Medicine, University of North Carolina Hospitals,
Chapel Hill, NC, USA

John Hunt, MD
Department of Pathology and Laboratory Medicine,
University of North Carolina Hospitals, Chapel Hill, NC, USA

Nathan D. Montgomery, MD, PhD
Department of Pathology and Laboratory Medicine,
University of North Carolina Hospitals, Chapel Hill, NC, USA

Chapter 1

Intraoperative Consultations of Tissue for Evaluation of Hematolymphoid Disorders or Malignancies: Appropriate Handling, Including Touch Preparation Cytology Versus Frozen Section

INTRODUCTION

Significant strides have been made in flow cytometric, cytogenetic, and molecular techniques in diagnosing and prognosticating hematolymphoid malignancies, as outlined in the recent World Health Organization (WHO) classification of tumors of hematopoietic and lymphoid tissues. Tissue biopsies acquired from patients with a suspected diagnosis (or prior history) of hematolymphoid malignancy should thus be handled in such a way that all necessary information may be gained for diagnostic and prognostic purposes. In general, if the biopsy is acquired from a "lymphoid" site (i.e., tonsil/adenoid, lymph node, thymus, spleen, etc.), the tissue should be handled as a "lymphomatous" specimen until proven otherwise, since these are irreplaceable specimens, and they should be triaged appropriately. If the biopsy is from an extranodal site and a diagnosis of a hematolymphoid malignancy is the primary concern, such biopsies should also be handled as a "lymphomatous" specimen. When in doubt, it is always better to process the sample as a "lymphomatous"

C.H. Dunphy (ed.), *Frozen Section Library: Lymph Nodes*,
Frozen Section Library, 10, DOI 10.1007/978-1-4614-1253-3_1,

specimen, to ensure that all appropriate testing has been instituted for diagnostic and prognostic purposes.

PROPER HANDLING OF "LYMPHOMATOUS" SPECIMENS

"Lymphomatous" specimens are ideally triaged for histologic sectioning, flow cytometric immunophenotyping (FCI), cytogenetic studies, and frozen tissue banking for possible future, indicated molecular studies. Since triaging of "lymphomatous" specimens is critical for diagnostic and prognostic purposes and intraoperative frozen sectioning of such specimens often compromises the tissue available for appropriate triaging and is fraught with misdiagnoses (to be discussed in future chapters), intraoperative frozen sections of such "lymphomatous" specimens is highly discouraged. The following protocol is highly recommended for handling such specimens.

Fresh tissues that come to the frozen section room with a history of a hematolymphoid malignancy, or labeled "suspicious for lymphoma," "for lymphoma workup," or similarly are accessioned and handled in the following way:

1. Lymph nodes, tonsils/adenoids, and thymus:
   (a) Perform touch preparations in the frozen section room.
   (b) If a nonhematolymphoid malignancy is identified, route to surgical pathology.
   (c) Otherwise, route to hematopathology for "lymphoma" workup.
   (d) If the TP is acellular or equivocal, one may either perform a frozen section and triage as above, or route to hematopathology directly, if tissue is compromised for appropriate triaging purposes. Every attempt should be made to triage these cases without freezing tissue, to ensure that adequate tissue is available for appropriate triaging purposes.

2. Extranodal or nonlymphoid tissues:
   (a) Perform touch preparations in the frozen section room.
   (b) If the touch preparation is interpreted as at least suspicious for a hematolymphoid malignancy, route to hematopathology.
   (c) Otherwise, route to surgical pathology.
   (d) If the TP is acellular or equivocal, perform a frozen section and triage as above.

      Every attempt should be made to triage these cases without freezing tissue.

      If frozen sections must be performed, one should limit the amount of tissue frozen.

Note: Nodular sclerosing Hodgkin lymphoma or other sclerotic lymphomas may be sparsely cellular on the touch preparation. Every attempt should be made to avoid freezing these specimens when these diagnoses are clinically suspected.

Once the specimen is routed to hematopathology, the tissue is triaged for diagnostic and prognostic purposes. Depending on the amount of tissue acquired, one may need to prioritize the triaging of the sample. Adequate, well-fixed, and well-sectioned tissue for histopathologic review is always the first priority, since histopathology remains the mainstay of diagnosis. The amount of tissue placed in a cassette should be less than the size of a nickel and no thicker than 3 mm. Ideally, tissues are fixed in non-buffered formalin to preserve immunoreactivity for possible indicated immunohistochemical staining and in B-5 fixative to enhance cytologic details. Depending on the amount of tissue and clinical history, tissue may then be triaged for FCI and/or cytogenetic studies. The nature of the tissue biopsy (i.e., non-sclerotic, cellularity, etc.) determines the amount of tissue necessary for FCI. A touch preparation should always be made of the portion of tissue to be considered for processing for FCI, to assess for suitability (i.e., to assess the presence of adequate tissue and the cell populations represented). Six touch preparations of the tissue are recommended – two of which may be Wright's stained to evaluate for adequacy of the specimen. The remaining unstained touch preparations may be refrigerated, and possibly used for future ancillary studies (i.e., indicated fluorescent in situ hybridization – FISH studies, etc.). Tissue is prepared as a single cell suspension for FCI by teasing the tissue in a small Petri dish with a scalpel against a fine mesh into Roswell Park Memorial Institute (RPMI) 1640 medium (Cellgro).

Suggested panels for FCI to evaluate lymphoid malignancies are recommended in Table 1.1.

TABLE 1.1 Recommended flow cytometric immunophenotypic panels to evaluate lymphoid malignancies.

Initial screen		
	Lymphoma	*Plasma cell*
	G1/G1/45/G1	G1/G1/45/G1
	G2/G2/45/G2	G2/G2/45/G2
	71/33/45/14	71/33/45/14
	10/5/45/23	3/138/45/56
	3/2/45/20	20/38/45/HLA-DR
	L/K/45/19	L/K/45/19

(continued)

TABLE 1.1 (continued)

	52/7/45/19	
	3/8/45/4	
	? Hematopoietic	
	3/19/45	
Follow-up		
Hairy Cell/MZL	*T cell*	*LGL*
22/11c/45	3/8/45/4	3/8/45/4
103/25/45		3/16 + 56/45/-
		57/-/45/56
	Adult T cell	
	7/25/45	*Plasma cell*
±CLL		3/138/45/56
20/34/45/38		20/38/45/HLA-DR
5/13/45/79b		
3/ZAP-7019		*ALCL*
		30/19/45/3

MZL marginal zone lymphoma, *LGL* large granular lymphocytosis/leukemia, *CLL* chronic lymphocytic leukemia, *ALCL* anaplastic large cell lymphoma

TABLE 1.2 Recommended flow cytometric immunophenotypic panels for leukemic processes.

Initial screen		
	G1/G1/45/G1	
	G2/G2/45/G2	
	71/33/45/14	
	10/19/45/20	
	34/56/45/117	
	7/64/45/HLA-DR	
	15/13/45/11b	
	5/2/45/3	
	L/K/45/19	
Follow-up		
B cell	*T cell*	*Myeloid*
22/24/45	3/8/45/4	36/GlyA/45
	3/1a/45	61/14/45

(continued)

TABLE 1.2 (continued)

	Intracellular
	3/MPO/45
	22/_/45
	TdT/_/45

Suggested panels for a leukemic process are recommended in Table 1.2.

If ample tissue is available after triaging appropriate tissue for FCI, a portion of tissue (at least 3 mm) should then be obtained and placed in RPMI media for routine cytogenetic studies. If additional tissue is available, it should ideally be frozen in liquid nitrogen and placed at –70° for possible future molecular studies.

Chapter 2

Applications of Touch Preparation Cytology to Intraoperative Consultations: Lymph Nodes and Extranodal Tissues for Evaluation of Hematolymphoid Disorders

INTRODUCTION

As discussed in Chap. 1, fresh tissues from "lymphoid" sites that come to the frozen section room with a history of a hematolymphoid malignancy, or labeled "suspicious for lymphoma," "for lymphoma workup," or similarly have touch imprints prepared and are Wright's stained. If a non-hematolymphoid malignancy is identified, the tissue is routed to surgical pathology. Otherwise, the tissue is routed to hematopathology. For extranodal tissues, if the touch preparation is interpreted as at least suspicious for a hematolymphoid malignancy, the tissue is routed to hematopathology. This chapter further discusses the applications of touch preparation cytology for the evaluation of such specimens in the intraoperative setting.

DETERMINATION OF A LYMPHOID PROCESS

Touch preparation cytology as an intraoperative consultation upon a specimen from a "lymphoid" site, being evaluated for a hematolymphoid malignancy, is primarily performed to determine the adequacy of the specimen and to confirm the presence of "lymphoid" tissue. Reactive lymph nodes typically demonstrate a polymorphic population, composed primarily of small lymphocytes intermingled

C.H. Dunphy (ed.), *Frozen Section Library: Lymph Nodes*,
Frozen Section Library, 10, DOI 10.1007/978-1-4614-1253-3_2,

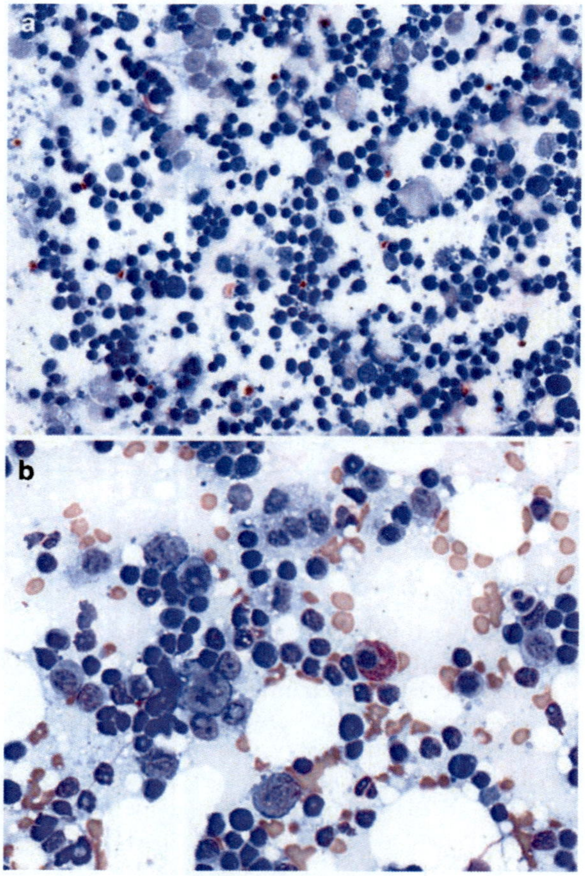

FIGURE 2.1 Touch preparation cytomorphology of reactive lymphoid process. (**a**) Polymorphous population composed predominantly of small lymphocytes with scattered larger forms and (**b**) an occasional polymorphonuclear cell and eosinophil.

with a mixture of intermediate-sized lymphocytes, scattered larger forms, occasional plasma cells, and histiocytes. Depending on the nature of the reactive process, occasional neutrophils and eosinophils may also be observed (see Fig. 2.1). A lymphomatous process, such as a non-Hodgkin lymphoma (NHL), may be suspected if the touch preparation demonstrates a cellular specimen composed of a monomorphous population of lymphoid cells (see Fig. 2.2). However, NHLs are a heterogeneous group of disorders, and some

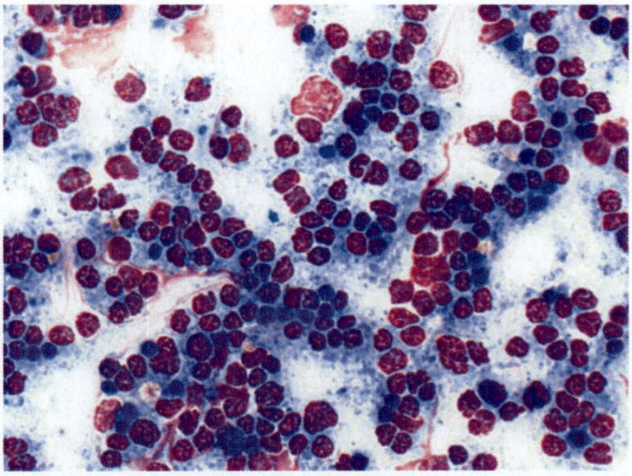

FIGURE 2.2 Touch preparation cytomorphology of non-Hodgkin malignant lymphoma NHL: Monotonous population of predominantly small lymphocytes in a case of small lymphocytic lymphoma.

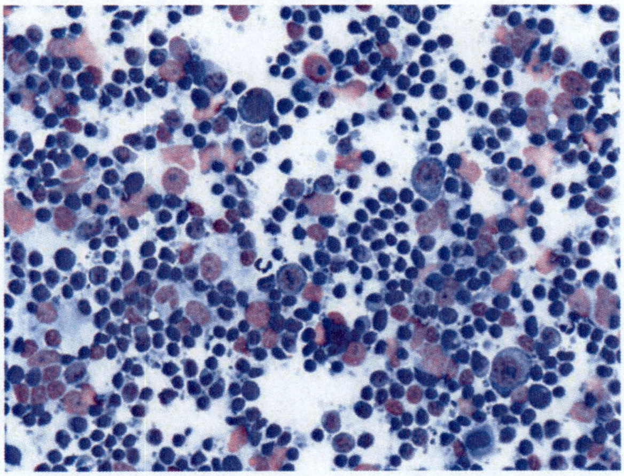

FIGURE 2.3 Touch preparation cytomorphology of lymphohistiocytic-rich large B-cell lymphoma: Rare scattered large abnormal lymphoid forms.

subtypes are composed of a polymorphous population. For example, in cases of T-cell- or lymphohistiocytic-rich large B-cell lymphoma, there is a background rich in reactive, non-neoplastic small lymphocytes and/or histiocytes with rare, scattered large abnormal lymphoid forms identified (see Fig. 2.3). A similar cytomorphology may also be seen in the lymphohistiocytic-rich variant of anaplastic large-cell lymphoma (ALCL). Likewise, the mixed cellularity subtype of classical Hodgkin lymphoma (cHL) often demonstrates a polymorphic population. However, in this subtype of cHL, one also usually observes abnormally enlarged, mono- and binucleated cells with prominent eosinophilic nucleoli (see Fig. 2.4). In addition,

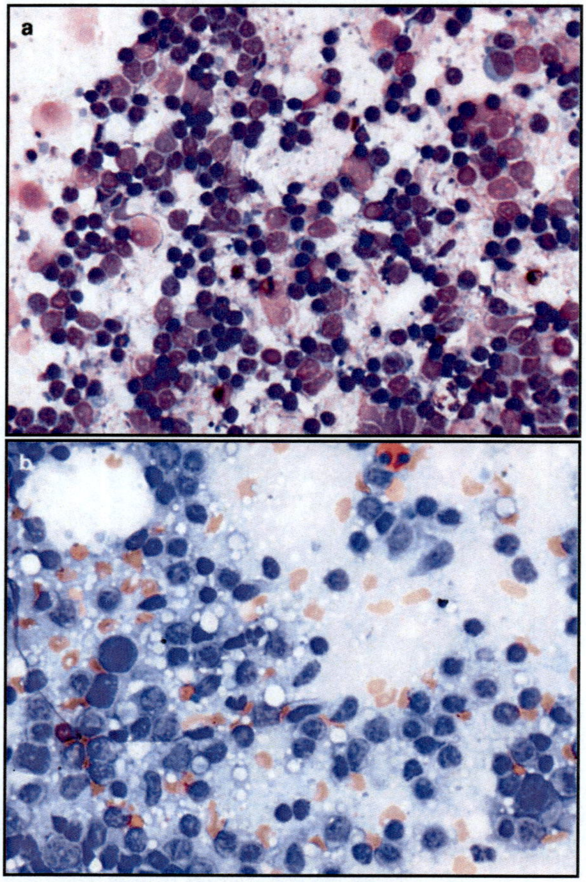

FIGURE 2.4 Touch preparation (TP) cytomorphology of mixed cellularity classical Hodgkin lymphoma (cHL). (**a**) Polymorphous population of predominantly small lymphocytes with scattered larger forms and smudged nuclei. (**b** and **c**) A different case demonstrating TP cytomorphology with scattered abnormal mononucleated and binucleated cells with very prominent nucleoli.

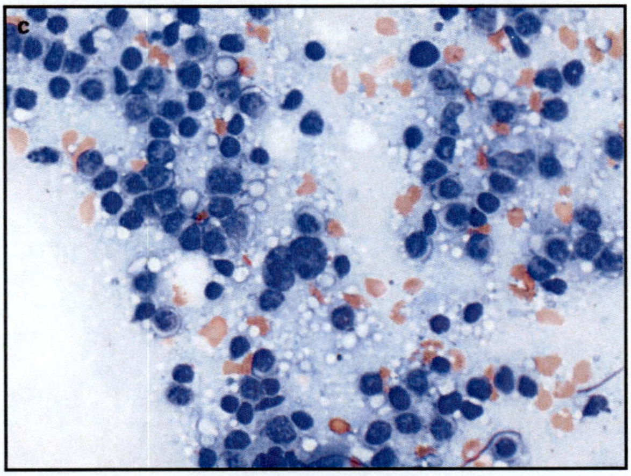

FIGURE 2.4 (continued)

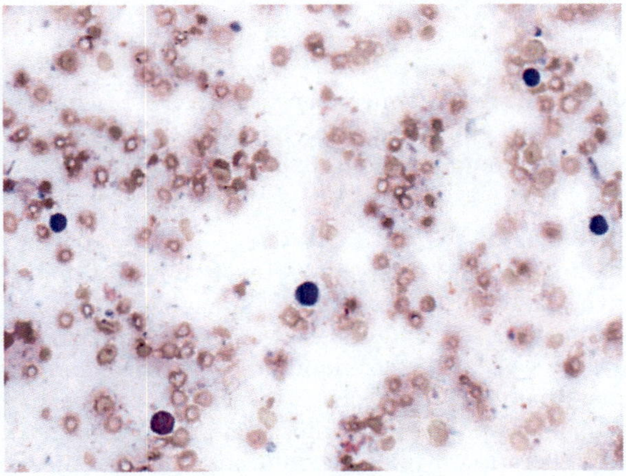

FIGURE 2.5 Touch preparation cytomorphology of the nodular sclerosing subtype of CHL: Markedly paucicellular touch preparation typical of this subtype, due to fibrosis.

some cases of lymphoma (i.e., the nodular sclerosing subtype of cHL and primary mediastinal B-cell lymphoma) are associated with marked sclerosis and yield a paucicellular touch preparation (see Figs. 2.5 and 2.6).

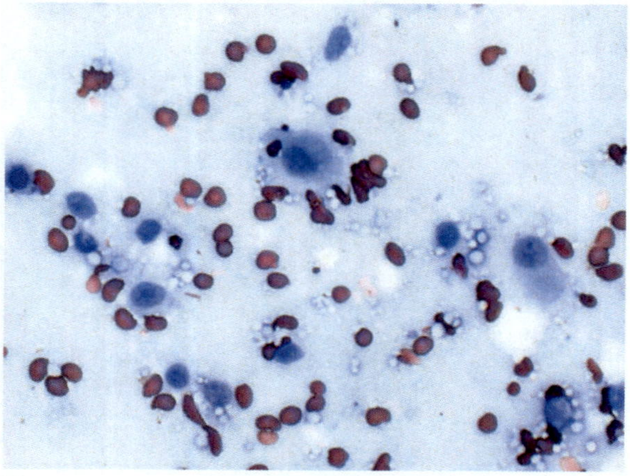

FIGURE 2.6 Touch preparation cytomorphology of the primary mediastinal large B-cell lymphoma: Markedly paucicellular touch preparation typical of this lymphoma, due to fibrosis.

In extranodal specimens, the primary purpose of the intraoperative touch preparation of a specimen suspected of a hematolymphoid malignancy is to confirm for adequacy of the specimen and to exclude a non-hematolymphoid malignancy.

SIGNIFICANCE OF GRANULOMATA

Granulomata may be observed in touch preparations of lymph nodes involved by granulomatous lymphadenitis (such as mycobacterial infections, sarcoidosis, etc.) and other reactive conditions (Fig. 2.7), as well as in eosinophilic granuloma and malignant lymphomas (both NHL and cHL). In triaging specimens intraoperatively for a lymphoma "workup," the presence of granulomata by touch preparation cytology should not dissuade one from considering a diagnosis of malignant lymphoma. In fact, granulomata are often seen in association with classical HL as well as in NHLs of B- and T-cell origin (see Fig. 2.8). In addition, the presence of abnormal lymphocytes may not always be appreciated since the granulomata may obscure the lymphomatous background in the touch preparations. For this reason, such specimens should be handled as a "lymphomatous" specimen and triaged appropriately. In general, when in doubt, it is better to triage as a lymphomatous specimen than to regret missing the opportunity to acquire essential diagnostic and prognostic data.

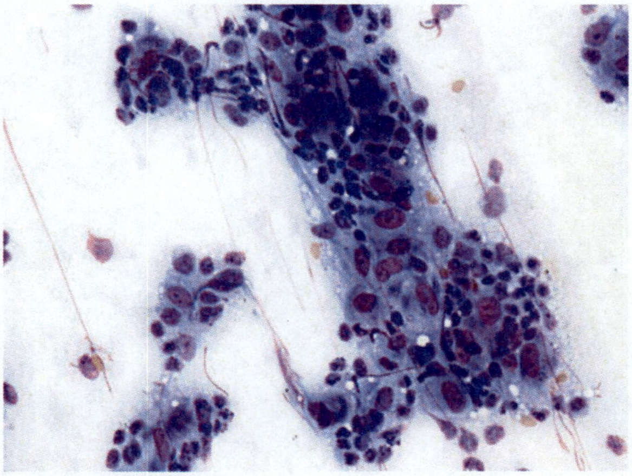

FIGURE 2.7 Touch preparation cytomorphology of granulomatous lymphadenitis: Numerous large cells with abundant cytoplasm representing numerous epithelioid histiocytes.

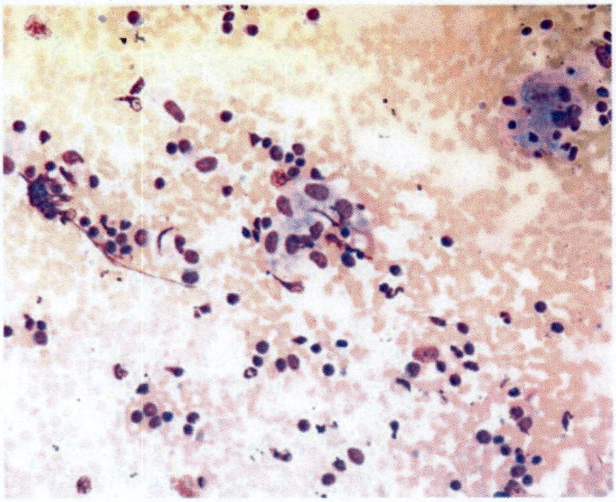

FIGURE 2.8 Touch preparation cytomorphology of a T-cell lymphoma associated with granulomata: Granuloma in the center with small lymphocytes (in the background) that do not appear particularly atypical. However, a subsequent biopsy revealed nuclear irregularities of the small lymphocytes associated with an aberrant flow cytometric immunophenotype and a T-cell clone by molecular studies. The presence of granulomata and lack of lymphocytic atypia on a TP does not exclude the possible presence of malignant lymphoma.

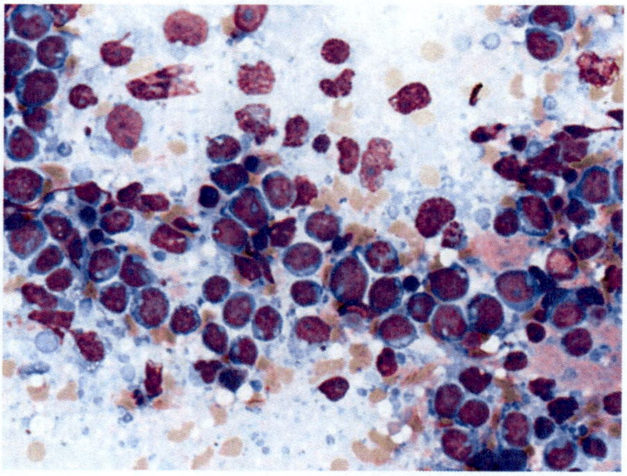

FIGURE 2.9 Touch preparation cytomorphology of diffuse large B-cell lymphoma: Discohesive, large abnormal lymphoid cells.

DIFFERENTIATION FROM NON-HEMATOLYMPHOID MALIGNANCY

Another primary purpose of the intraoperative evaluation of specimens with a suspected hematolymphoid malignancy is to exclude a non-hematolymphoid metastatic malignancy (i.e., melanoma, carcinoma, etc., in nodal sites) or a metastatic or primary malignancy in other lymphoid (i.e., tonsil) and extranodal sites. If a non-hematolymphoid malignancy is identified by TP cytology, the specimen is routed to surgical pathology instead of hematopathology. A malignant lymphoma is typically characterized by the presence of discohesive malignant cells (Fig. 2.9), whereas, non-hematolymphoid malignancies, such as carcinoma, are often characterized by the presence of cohesive sheets or clusters of malignant cells, occasional glandular formation, and/or molding of tumor cells (Fig. 2.10). However, poorly differentiated carcinomas as well as small round blue-cell tumors, neuroendocrine carcinomas, sarcomas, melanoma, and other non-hematolymphoid malignancies may mimic malignant lymphoma of various subtypes (i.e., ALCL, diffuse large-cell lymphoma, etc.) (see Figs. 2.11 and 2.12). In addition, ALCL may also show varying morphologies, including spindled forms as well as forms mimicking carcinoma with clustering of tumor cells (see Fig. 2.13).

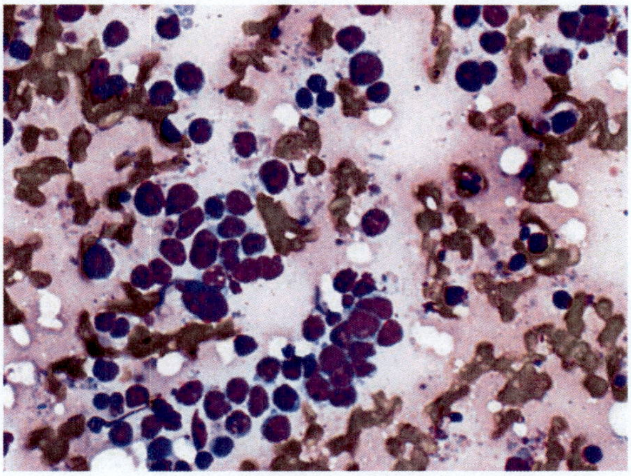

FIGURE 2.10 Touch preparation cytomorphology of metastatic carcinoma in lymph node: Molding of tumor cells.

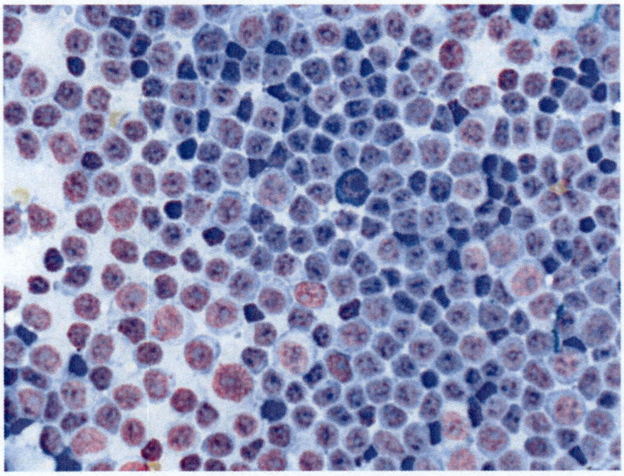

FIGURE 2.11 Touch preparation cytomorphology of metastatic melanoma to lymph node: Mimics malignant lymphoma.

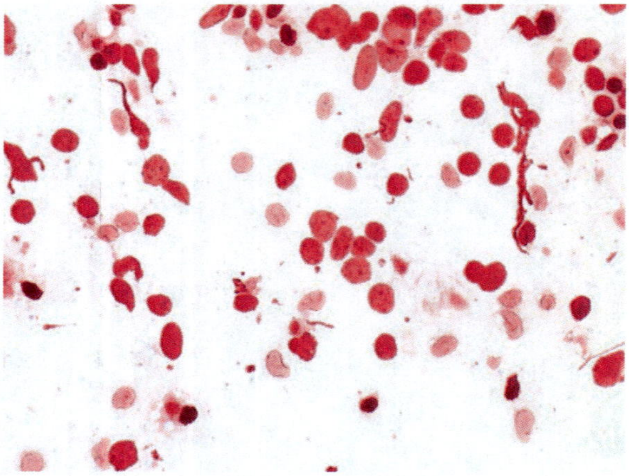

FIGURE 2.12 Touch preparation cytomorphology of neuroendocrine carcinoma: Mimics malignant lymphoma with large, discohesive lymphoid-appearing cells.

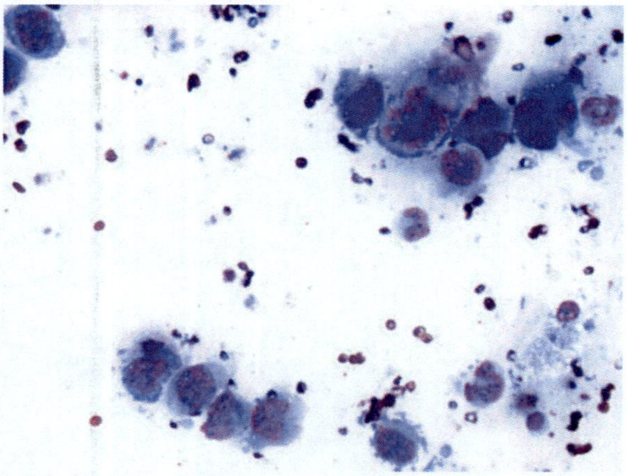

FIGURE 2.13 Touch preparation cytomorphology of anaplastic large cell: Mimics carcinoma with very large malignant cells and lining up of tumor cells.

HEMATOLYMPHOID MALIGNANCIES

Now the discussion focuses on hematolymphoid malignancies and what may be gleaned from touch preparation cytology. In general, leukemias and lymphomas are characterized by discohesive cytopathology.

Myeloid Sarcomas

In bone marrow specimens (i.e., bone marrow aspirate smears and touch preparations of bone marrow core biopsies), cytopathologic features will obviously depend on the clinical type of leukemia (acute or chronic) and the cell of origin subtype (myeloid or lymphoid). Acute leukemias are characterized by an abnormal increase in blasts, which are characterized by nuclei with fine, open chromatin, and varying numbers of nucleoli. Acute myeloid leukemias (AMLs) may show small lymphoid-appearing blasts in AML, M0 and larger blasts with more cytoplasm, and increasing granularity in AML, M1 and AML, M2. Acute promyelocytic leukemia (APL) has hypergranular and hypogranular variants. The hypergranular variant is characterized by abnormal promyelocytes with markedly granular and abundant cytoplasm. The hypogranular variant of APL may mimic acute monocytic leukemia (AML, M5b). True Auer rods help in distinguishing AML from lymphoblastic leukemia. Monoblasts are characterized by abundant bluish gray cytoplasm often with numerous cytoplasmic vacuoles. Pure erythroid leukemia (AML, M6b) demonstrates malignant, bizarre erythroblasts, and acute megakaryoblastic leukemia (AML, M7), megarkaryoblasts characterized by cytoplasmic blebs.

Since BM specimens are not typically evaluated intraoperatively, it is more likely that tissue forms of acute leukemia may be encountered in these situations. Tissue forms of AML have been described as myeloid sarcoma, monocytic sarcoma, and rarely erythroid sarcoma. Myeloid sarcomas may be composed of varying numbers of myeloblasts and more maturing myeloid elements, as well as eosinophils (see Fig. 2.14). Monocytic sarcomas are composed of sheets of blasts of monocytic lineage, which may be confirmed by enzyme cytochemical staining with nonspecific esterases (see Fig. 2.15). Erythroid sarcoma is composed of sheets of malignant erythroblasts, characterized by bizarre forms with dark blue cytoplasm containing varying numbers of vacuoles (see Fig. 2.16). Such sarcomas are considered tissue forms of AML. Large discohesive cells with cytoplasmic vacuoles, as seen in some types of AML, may be difficult to be distinguished cytomorphologically from large-cell lymphomas and Burkitt lymphoma, which may also have overlapping cytomorphological features (see Figs. 2.17 and 2.18). In addition, one must also keep in mind that the tissue form of lymphoblastic leukemia may present as lymphoblastic lymphoma, more frequently of precursor T-cell origin, but also rarely of precursor B-cell origin (see Figs. 2.19 and 2.20).

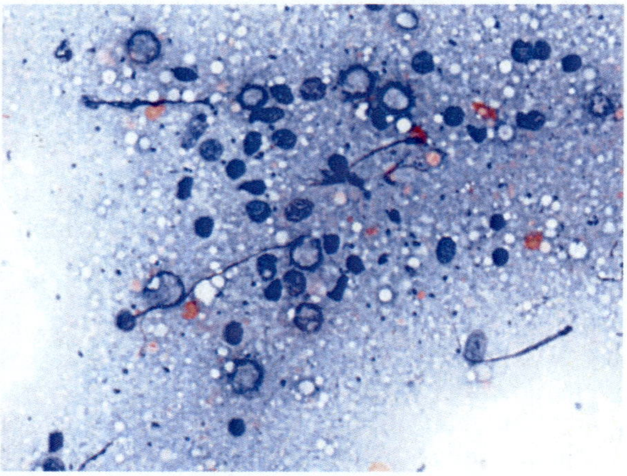

FIGURE 2.14 Touch preparation cytomorphology of myeloid sarcoma: Occasional blasts are seen with varying amounts of cytoplasm; there are also numerous stripped nuclei and abundant debris in the background.

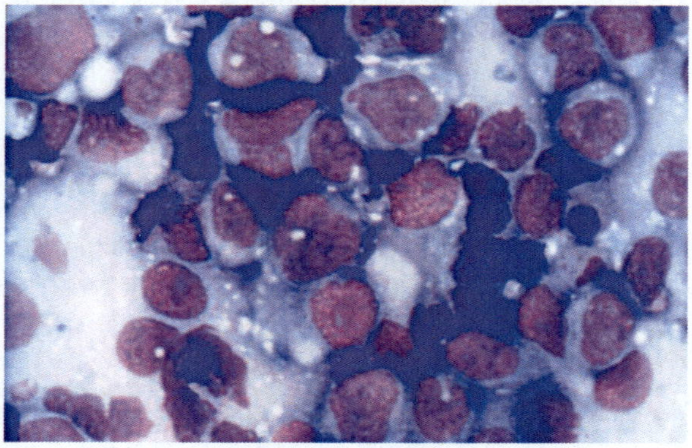

FIGURE 2.15 Touch preparation cytomorphology of monocytic sarcoma: Sheets of monoblasts.

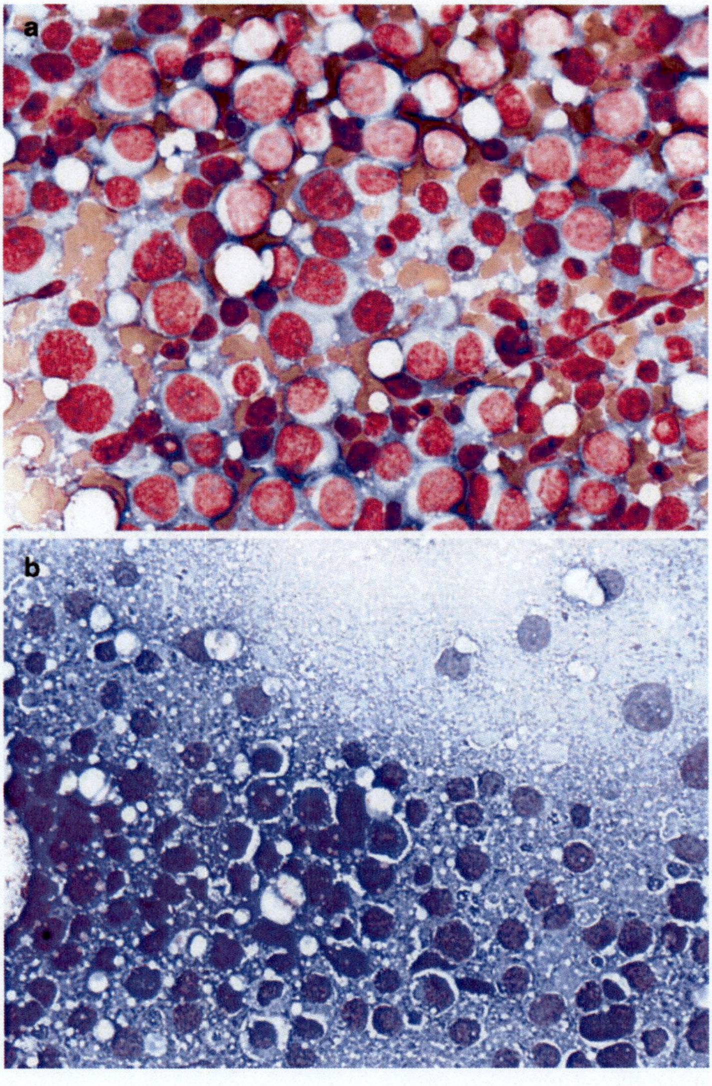

FIGURE 2.16 Touch preparation cytomorphology of erythroid sarcoma. (**a**) The blasts have associated light bluish cytoplasm in this view and (**b**) darker blue cytoplasm with apparent vacuoles in this view.

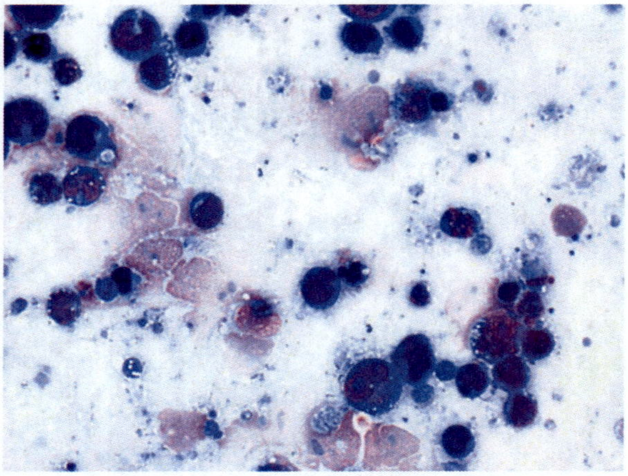

FIGURE 2.17 Touch preparation cytomorphology of diffuse large B-cell lymphoma: Cytoplasmic vacuoles.

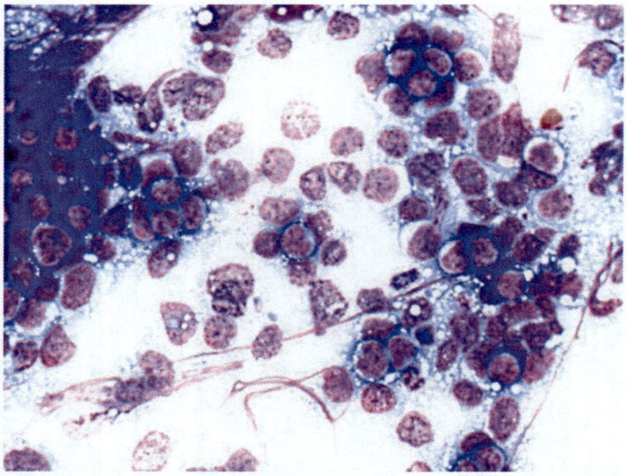

FIGURE 2.18 Touch preparation cytomorphology of Burkitt lymphoma: Cytoplasmic vacuoles.

Malignant Lymphomas Composed Predominantly of Small Lymphocytes

Small lymphocytic lymphoma, SLL; mantle cell lymphoma, MCL; follicular lymphoma, FL; lymphoplasmacytic lymphoma, LPL; marginal zone B-cell lymphoma, MZL

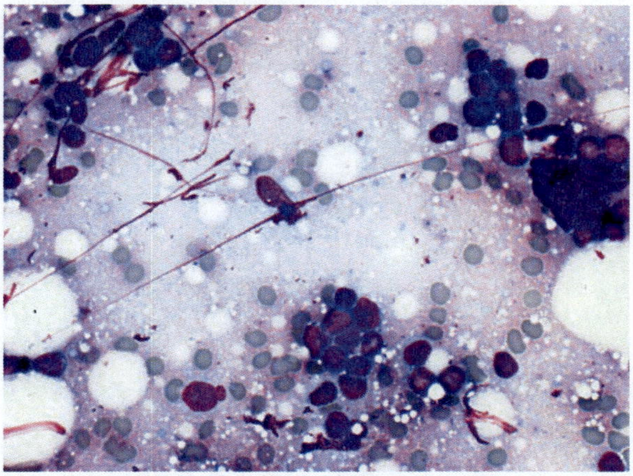

FIGURE 2.19 Touch preparation cytomorphology of T-lymphoblastic lymphoma: Clusters of lymphoblasts.

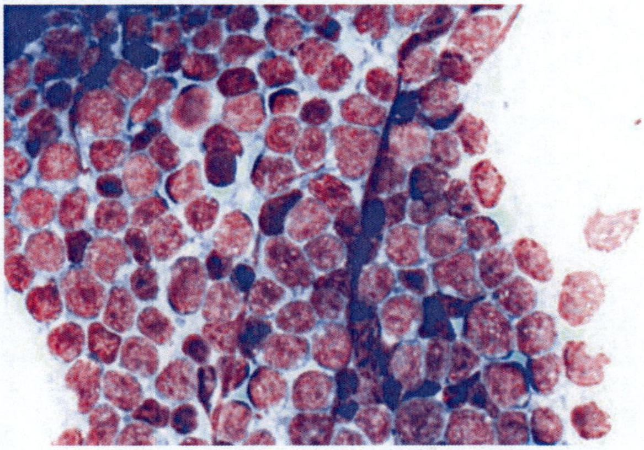

FIGURE 2.20 Touch preparation cytomorphology of B-lymphoblastic lymphoma: Sheets of lymphoblasts.

One can evaluate touch preparations of lymph nodes for the cytomorphological features characteristic of the NHLs composed predominantly of small lymphocytes as listed in Table 2.1. However, even with the recognition of these characteristic cytomorphological features, limitations are encountered in TP cytopathology as

TABLE 2.1 Characteristic cytomorphological features of the non-Hodgkin lymphomas composed predominantly of small lymphocytes.

Type of non-Hodgkin lymphoma	Cytomorphological features	Limitations
Small lymphocytic lymphoma	Monotonous population of predominantly small, round, and mature lymphocytes; prolymphocytes and paraimmunoblasts may be increased particularly in proliferation centers	Difficult to evaluate for large-cell transformation if, for example, touch preparation is from proliferation center
Mantle cell lymphoma	Monotonous population of predominantly small, slightly irregular lymphocytes	May not detect the presence of pink histiocytes in TP, which may aid in distinguishing this lymphoma from other types of NHL
Follicular lymphoma	Monotonous population of small, cleaved lymphocytes	Cannot exclude significant large-cell component; reactive germinal centers may be difficult to distinguish from grade 2 follicular lymphoma; cannot determine pattern of involvement
Marginal zone B-cell lymphoma	Monotonous population of monocytoid-appearing lymphocytes, characterized by small lymphoid nuclei associated with abundant clear cytoplasm, and intermixed with varying numbers of plasma cells	May be difficult to distinguish from lymphoplasmacytic lymphoma
Lymphoplasmacytic lymphoma	Population composed of an intimate mixture of small, round lymphocytes, plasmacytoid lymphocytes, and mature plasma cells	May be difficult to distinguish from marginal zone B-cell lymphoma

also listed in this table and are inherent to this technique, such as possible sampling issues and the inability to evaluate for the pattern of involvement.

Determination of the Proportion of Large Cells and Pattern of Involvement

As alluded to in Table 2.1, touch preparation cytology does not allow for a definitive determination of the proportion of large cells, for example, in follicular lymphoma, SLL, MCL, and other types of malignant lymphoma, in which the proportion of large cells is important in the correct subclassification and treatment approach. Likewise, TP cytology does not allow for the determination of the pattern of involvement by a malignant lymphoma, which may also be important in establishing an accurate diagnosis.

Evaluation of Various Subtypes of Large-Cell Lymphoma

Diffuse large B-cell lymphoma, DLBCL; blastic mantle cell lymphoma, etc.

Touch preparation cytology may reveal a cellular sample composed of sheets of large discohesive lymphoid-appearing cells, which is consistent with a diffuse large-cell lymphoma, but histologic sectioning is still necessary to identify features that may be important diagnostically, prognostically, and for therapeutic options. For example, a background of follicular lymphoma or coexistent cHL cannot be determined by TP cytology alone. In addition, even the best TP cytology does not allow for the distinction between DLBCL, blastic mantle cell lymphoma, and other forms of diffuse lymphoma composed of large lymphoid cells, as well as from lymphoblastic lymphoma.

Differentiation of Diffuse Large B-Cell Lymphoma from Burkitt Lymphoma (BL) and B-Cell Lymphoma, Unclassifiable with Features Intermediate Between DLBCL and BL

Similarly, it may be difficult or impossible by TP cytology alone to distinguish between DLBCL, Burkitt lymphoma, and those B-cell lymphomas, unclassifiable with features intermediate between DLBCL and Burkitt lymphoma. Classical Burkitt lymphoma is composed of a monotonous population of medium-sized lymphocytes associated with dark blue cytoplasm containing multiple cytoplasmic vacuoles. There is associated abundant apoptotic debris. When classical cytopathology is identified in association with a c-myc translocation, a diagnosis of BL may be readily applied. However, some cases may not show the typical cytopathologic findings and be difficult to distinguish from the other entities described above.

Identification of Anaplastic Large-Cell Lymphoma

Likewise, the common variant of ALCL is composed of numerous "hallmark" cells, characterized by a doughnut shape and multinucleated forms. When such classical cytopathology is identified in association with an ALK translocation, a diagnosis of ALCL may be readily applied. However, ALCL is known to be a great histologic mimicker (similar to malignant melanoma in this regard) with many cytomorphological and histological variants. These variants may mimic carcinoma, sarcoma, other malignancies, and even reactive proliferations. In such variant cases, a diagnosis of ALCL may be difficult to distinguish from other types of large-cell lymphoma, as well as from non-hematolymphoid malignancies and even reactive lymphoid proliferations.

Lymphomas with Marked Sclerosis

Classical Hodgkin lymphoma, primary mediastinal large B-cell lymphoma

As mentioned previously, lymphomas associated with marked sclerosis may yield an extremely paucicellular touch preparation for cytomorphological examination. These lymphomas often present in extranodal sites, such as the mediastinum. One should thus be aware of this issue and consider such lymphomas a diagnostic possibility for triaging purposes, unless a non-hematolymphoid malignancy is identified.

Differentiation of Thymoma, Thymic Hyperplasia, and Ectopic Thymus from T-Lymphoblastic Lymphoma

By flow cytometry, the immunophenotype of thymic tissue (i.e., thymoma, thymic hyperplasia, and ectopic thymus) may be identical to T-lymphoblastic lymphoma (T-LL, i.e., the common thymocyte stage immunophenotype). However, in most cases, one should be able to cytomorphologically distinguish benign thymic tissue or thymoma from T-LL, since T-LL is characterized by sheets of lymphoblasts (see Fig. 2.21). However, some cases may be morphologically challenging, especially in TP evaluations. Fortunately, there are differences in the flow cytometric immunophenotype of thymic tissue and T-LL, which may aid in morphologically challenging cases. T-LL shows a tight expression pattern of CD3 and TdT and CD3 vs. CD4, whereas, thymic tissue shows a heterogeneous expression pattern of these same markers (see Figs. 2.22 and 2.23). For such reasons, mediastinal tissues with a possible diagnosis of thymic hyperplasia, thymoma, or T-LL should be evaluated by hematopathology to triage the specimen appropriately for ancillary testing, which may

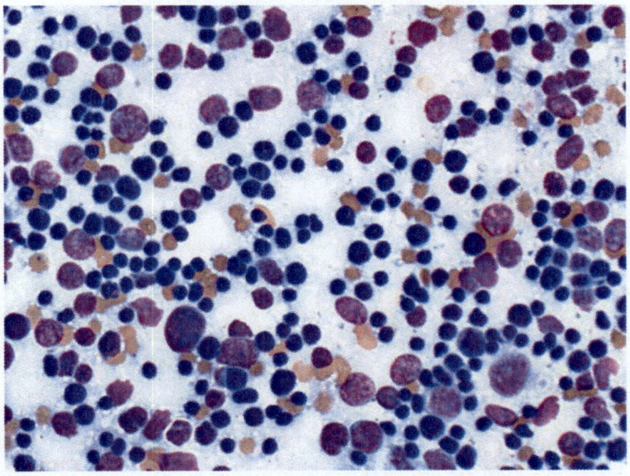

FIGURE 2.21 Touch preparation cytomorphology of thymoma: Polymorphous population of small lymphocytes with scattered larger forms and epithelioid cells.

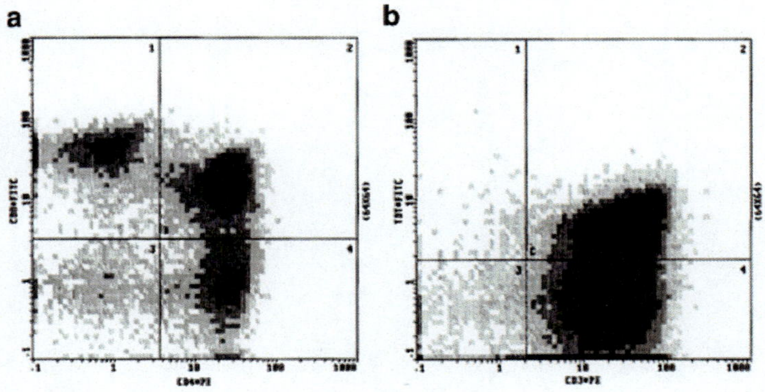

FIGURE 2.22 Flow cytograms of thymoma. (**a**) Smear pattern of CD4 (x-axis)/CD8 (y-axis) characteristic of thymoma and (**b**) smear pattern of TdT (y-axis)/CD3 (x-axis) characteristic of thymoma.

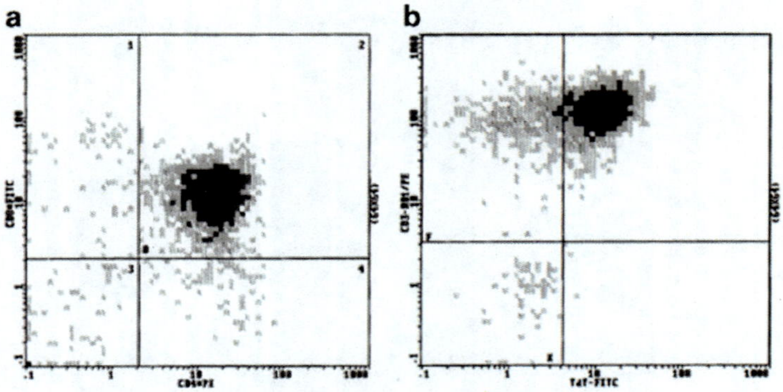

FIGURE 2.23 Flow cytograms of T-lymphoblastic lymphoma. (**a**) Tight pattern of CD4 (*x*-axis)/CD8 (*y*-axis) characteristic of T-lymphoblastic lymphoma and (**b**) tight pattern of TdT (*x*-axis)/CD3 (*y*-axis) characteristic of T-lymphoblastic lymphoma.

be critical in establishing an accurate diagnosis in morphologically challenging cases. Likewise, in non-mediastinal tissues, ectopic thymus may be difficult to distinguish from other lymphoid lesions by cytopathology alone.

Chapter 3
Applications of Frozen Section to Intraoperative Evaluations of Extranodal Tissues for Hematolymphoid Disorders and Diagnostic Pitfalls of Frozen Section to Intraoperative Evaluations of Lymph Nodes for Hematolymphoid Disorders

INTRODUCTION

Since frozen sections are preferably not routinely performed on lymph node specimens for hematolymphoid malignancies, this chapter is primarily devoted to the applications of frozen section to extranodal tissues being evaluated for hematolymphoid disorders. The chapter also highlights artifactual changes and diagnostic pitfalls of frozen sections in lymph nodes and "lymphoid" tissues, further confirming the limited usefulness of frozen sections in lymph nodes and "lymphoid" tissues suspected of a hematolymphoid malignancy.

As discussed previously in Chap. 1, extranodal tissues with a suspected diagnosis of a hematolymphoid disorder are generally treated as "lymphoid" tissues. However, if the touch preparation is acellular or equivocal, a frozen section may be performed. One should, of course, be aware that nodular sclerosing classical Hodgkin

C.H. Dunphy (ed.), *Frozen Section Library: Lymph Nodes*,
Frozen Section Library, 10, DOI 10.1007/978-1-4614-1253-3_3,

lymphoma (NS cHL), or other sclerotic lymphomas, may be sparsely cellular on the touch preparation. For such reasons, every attempt should be made to avoid freezing these specimens when these diagnoses are clinically suspected, to ensure adequate tissue is available for ancillary testing, as discussed previously. However, in some situations, it may be necessary to perform a frozen section. In such situations, one should limit the amount of tissue frozen.

INDICATIONS FOR LIMITED FROZEN SECTIONS OF EXTRANODAL TISSUE BIOPSIES

The following are indications for limited frozen sectioning of extranodal tissue biopsies suspected of a hematolymphoid malignancy:

- Markedly paucicellular specimen on touch preparation cytology with ample specimen for permanent histologic sections (i.e., for histologic evaluation and possible extensive immunohistochemical staining) as well as for cytogenetic and possible molecular studies. It is understood that the specimen may not yield adequate cells for flow cytometric analysis.
- Patient has a history of a nonhematolymphoid malignancy.
- Patient has a suspected equal likelihood of a nonhematolymphoid malignancy.

FROZEN SECTION ARTIFACTS AND DIAGNOSTIC PITFALLS

Since frozen sections are not routinely performed on "lymphoid" tissues suspected of a hematolymphoid malignancy, we will now discuss examples of artifactual changes due to frozen sectioning and diagnostic pitfalls that may or have been encountered in frozen sections of "lymphoid" tissues. These artifactual changes and diagnostic pitfalls further confirm the limited usefulness of frozen sections in "lymphoid" tissues.

Artifactual Changes

Several factors may lead to artifactual changes due to the frozen section technique occurring at any stage of the process – from the actual surgical biopsy of the specimen to the final placement of the coverslip. Obviously, electrosurgery will often lead to polarization of cells with a characteristic streaming appearance. Cauterization of the specimen may result in coagulation of proteins with decreased cellular detail and an amorphous appearance [1]. Although these techniques are not often used in lymph node biopsies, they would also result in artifactual changes in the permanent tissue sections. Likewise, dye and tissue "nicks" are not used in marking lymph nodes for tissue orientation or for identification of margins, so these techniques do not typically impose any additional artifacts in lymph node specimens, unless the specimen is initially considered a nonhematolymphoid malignancy. One of the main reasons for performing frozen sectioning on a tumor is for margins, which simply does not apply in hematolymphoid malignancies.

The following is a list of steps during the actual frozen section process that may introduce subsequent artifactual changes:

1. Proper temperature of cryostat: Over- and under-freezing may lead to artifacts. If the temperature is warmer than −20°C, the tissue will not cut evenly. In these instances, it will compress upon itself, leaving a tissue "blob." Over-freezing, due to an improperly set cryostat or excessive use of cryoquick or liquid nitrogen, may cause shattering of tissue.
2. Proper securement of chuck and blade in cryostat: Knife chatter occurs when the chuck is not securely attached to the cutting arm or when the blade is not securely attached. This causes the tissue to move closer and further from the blade, leading to thicker and thinner cuts of tissue on the same section.
3. Selection and size of tissue: In selecting lymph node tissue for frozen sectioning, one must ensure first that enough tissue will remain for all appropriate ancillary studies [i.e., permanent histologic sections ideally at least one section in formalin (for immunohistochemistry) and one in B5 (for cellular detail), flow cytometric immunophenotyping, cytogenetic studies, and cryopreservation (for possible molecular studies)]. Immunohistochemistry is ideally performed on formalin-fixed, paraffin-embedded tissue that has not previously undergone frozen sectioning. Even when abundant tissue is available for all studies, one should avoid areas of calcification or possible boney areas, which may cause shattering of the tissue upon frozen sectioning. Grossly suspicious areas for hematolymphoid malignancy are those that have a light pink to tan, fish-flesh appearance. However, even in ideal situations, frozen section evaluation of lymph node specimens is not encouraged and may be fraught with diagnostic challenges due to sampling errors and other artifactual changes.
4. Mounting of tissue: Properly mounting tissue onto the chuck is an important step in tissue processing. If the specimen is not centrally placed on the chuck with a proper amount of mounting medium, or if the specimen is uneven with undulations (i.e., the tissue does not have a smooth surface or is not sufficiently flattened), frozen sectioning may demonstrate "skip" areas on the slides and folding of tissue upon itself.
5. Slide preparation: When preparing slides, tissue folds should be avoided. It is important to keep the tissue flat on the cryostat

blade and to maintain this as it is transferred to the warm glass slide [2, 3].

6. Staining of slide: Even and consistent staining is very important for proper interpretation. Using a fixative such as formalin, acetone, or alcohol will improve slide quality. Stains need to be periodically changed, as some will "ripen" over time and others may "weaken." In addition, during the staining process, folding may occur as the slides are dipped in and out of the staining baths. Moisture on the slide or underneath the tissue may also cause folding. Slides with an adherent coating that holds a positive charge are available. These may help minimize folding of tissue by attracting tissue to the slide.
7. Coverslip application: Air bubbles caused by improper coverslip application are common and may obscure underlying tissue. Care must be taken to use an adequate amount of mounting medium when applying the coverslip. The coverslip should be replaced if bubbles occur.

Due to these artifactual changes that may be introduced, the frozen sectioning of "lymphoid" tissue often results in "unreal" changes in the cellular morphology and architectural relationships of "lymphoid" tissue, which are critical in an accurate, reliable diagnosis.

Diagnostic Pitfalls

The following discussion includes the diagnostic dilemmas that have been described in frozen sections of lymphoid tissues, based on such artifacts encountered during frozen sectioning.

Distinguishing Reactive or Benign Lymphoid Proliferations from Malignant Lymphoma

Due to artifactual changes in cellular detail, it may be quite difficult to distinguish follicular lymphoma, for example, from follicular hyperplasia. In addition, artifactual changes in architectural relationships may impart a false impression of a follicular pattern, making it difficult to even accurately assign a follicular growth pattern in lymphoid tissue (see Figs. 3.1 and 3.2).

Distinguishing Thymoma from Malignant Lymphoma

A case of classical Hodgkin lymphoma (cHL) has been described, which was diagnosed as thymoma on frozen section [4]. The diagnosis of mediastinal cHL is one of the most difficult to ascertain on small biopsies. The lesion contains variable amounts of fibrosis. This fibrosis, which is usually nodular, may simulate the fibrous septae which lobulate a thymoma. These fibrous septae are generally more sharply delineated and often have a more angular shape

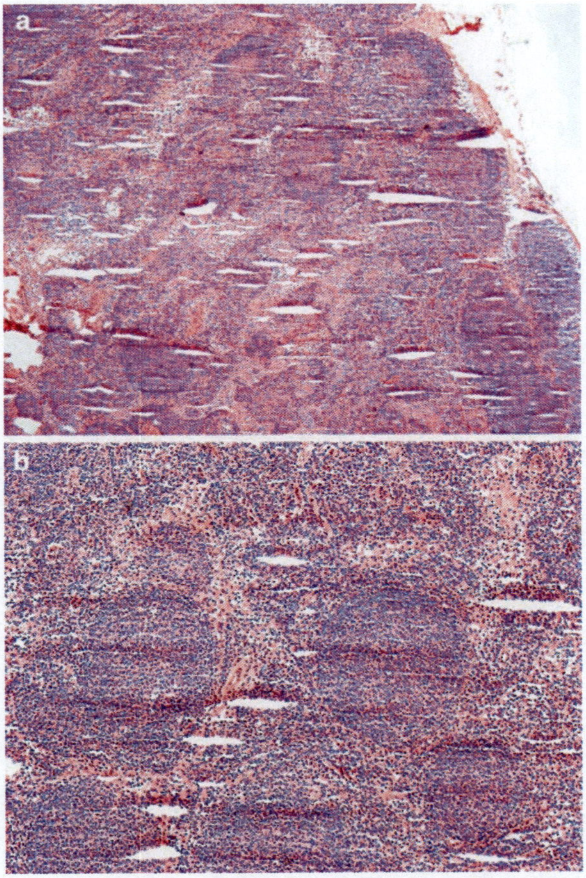

FIGURE 3.1 Frozen section of follicular lymphoma: Mimics "follicular hyperplasia" with open sinuses and artifactual polarization of malignant follicles [(**a**) low power; (**b**) higher power].

than the nodular fibrosis of cHL. However, these differences may be difficult to appreciate on frozen sections of small biopsies. Furthermore, both thymomas and mediastinal cHL have a relatively slow growth rate, may be associated with no symptoms or with nonspecific symptoms, and thus may be clinically confused. Therefore, the diagnosis of thymoma must be made with caution on a frozen section, since a false diagnosis may lead to a useless and possibly aggressive resection of an infiltrative lesion of cHL. The characteristic fibrosis in cHL (especially at the periphery of

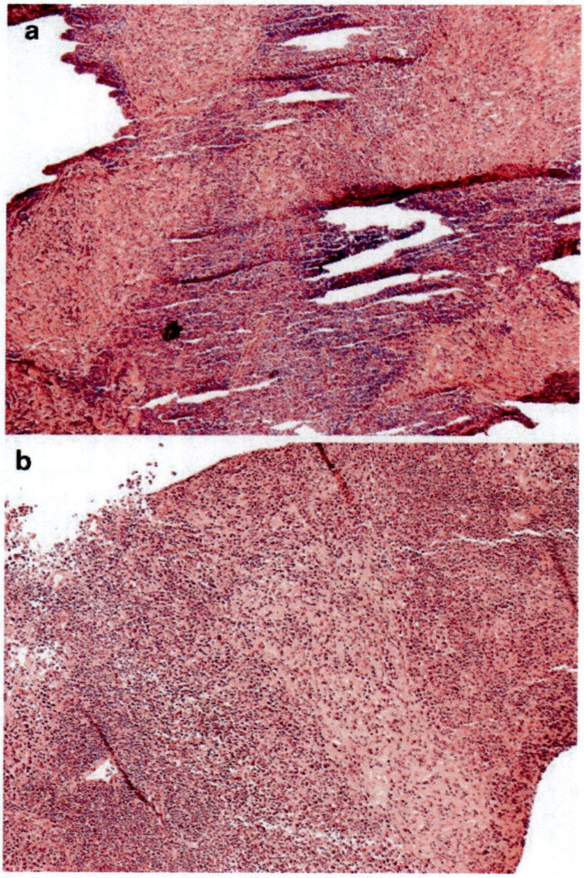

FIGURE 3.2 Frozen section of diffuse large B-cell lymphoma. (**a**) 0 artifactual "nodular" pattern due to sclerosis, shattering of tissue during the frozen section process, and loss of cellular detail; (**b**) permanent section reveals sheets of large malignant cells in the sclerosis.

the tumor) may also be very important in establishing a diagnosis. Thus, to avoid the possibility of a wrong diagnosis, such as fibrous mediastinitis, samples must be sufficiently large and contain cellular material.

Likewise, it may be difficult to distinguish thymoma from T-lymphoblastic lymphoma based on frozen sectioning alone. There may be artifactual changes resulting in the loss of cellular detail and architectural relationships, which may make this distinction quite difficult (see Fig. 2.21 in Chap. 2).

Distinguishing Malignant Lymphoma with Necrosis from an Infectious Process

In the same report by de Montpreville et al. [4], a false frozen section diagnosis of "malignant lymphoma and associated tumor necrosis" led to the absence of an important bacteriologic study in the case of an infectious pseudotumor of the anterior mediastinum, secondary to a chest cutaneous staphylococcal infection.

Likewise, there have been misdiagnoses by frozen section of "granulomatous lymphadenitis" in an actual case of NS cHL and of "abscess" in an actual case of recurrent cHL (see Figs. 3.3 and 3.4).

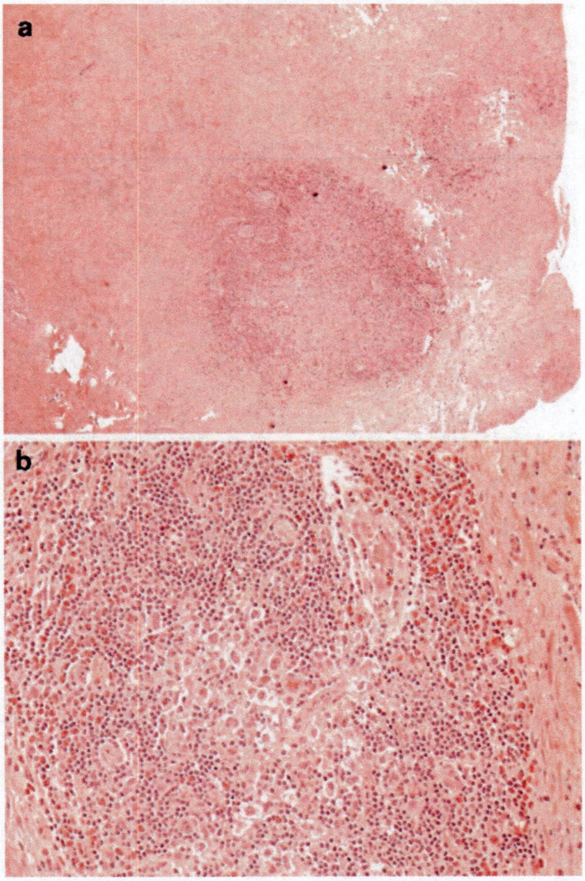

FIGURE 3.3 Frozen section of nodular sclerosing classical Hodgkin lymphoma: Misdiagnosed as "granulomatous lymphadenitis" due to nodular pattern of granulomata and distortion of nuclear detail [(**a**) low power; (**b**) higher power].

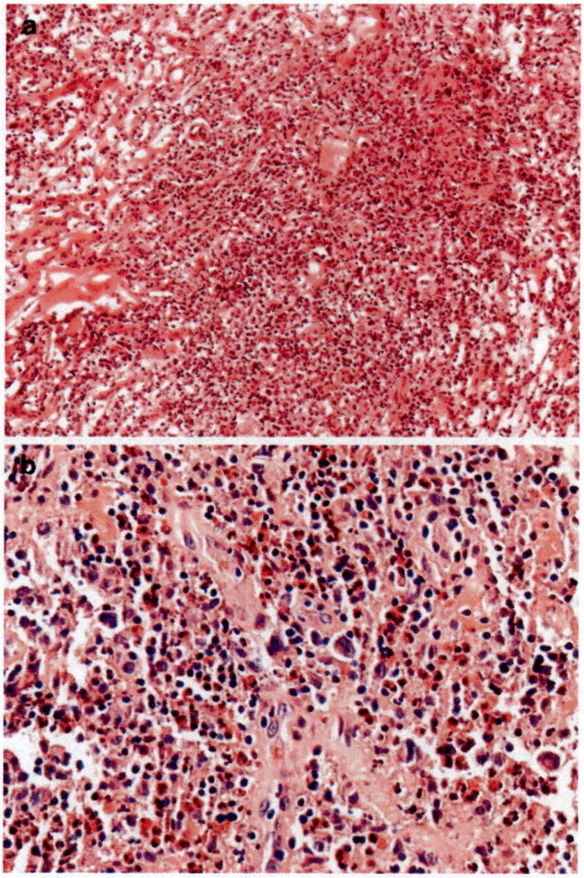

FIGURE 3.4 (**a**) Frozen section of recurrent classical Hodgkin lymphoma: Misdiagnosed as "abscess" due to numerous eosinophils with distorted nuclear detail and appearance as neutrophils on frozen section; (**b**) permanent section reveals the clear presence of numerous eosinophils (not neutrophils) and scattered abnormal mononucleated cells, representing Reed–Sternberg variants.

In the first case (Fig. 3.3), artifactual changes of the frozen section made it difficult to distinguish histiocytes in the granulomata from mononucleated Reed–Sternberg (RS) cells. In the second case (Fig. 3.4), the abundance of neutrophils obscured the rare, scattered RS cells on the frozen section.

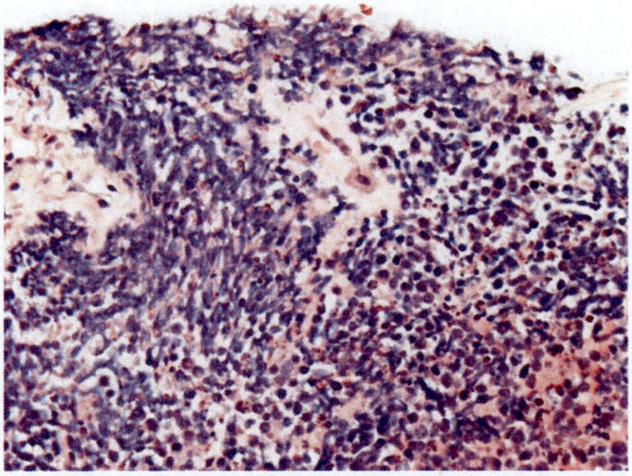

FIGURE 3.5 Frozen section of small-cell carcinoma: Mimics malignant lymphoma with dyscohesive lymphoid-appearing cells.

Distinguishing a Hematolymphoid Disorder or Malignancy from Nonhematolymphoid Malignancy

In addition to frozen sections being fraught with misdiagnoses in lymphoid tissue due to artifactual changes, frozen sectioning is also less effective for a precise diagnosis of some primary mediastinal lesions, which may have close histologic appearances. There has been a description of an invasive thymoma that was resected during a second operation after a frozen section diagnosis of carcinoma [4]. Likewise, small-cell carcinoma may be confused with malignant lymphoma. On frozen section, the tumor cells may resemble lymphocytes and the lesion may be confused with a malignant lymphoma (see Fig. 3.5). Taking advantage of the anesthesia, one may then perform an unnecessary bone marrow biopsy. The resemblance may even lead to confusion with small crushed fragments of normal or hyperplastic lymph node. In mediastinal and non-mediastinal sites, small, round, blue-cell tumors may also be confused with malignant lymphoma (see Fig. 3.6). Likewise, it may be quite difficult to distinguish metastatic carcinoma from anaplastic large-cell lymphoma, which is often known to demonstrate a cohesive, paracortical, and/or sinusoidal growth pattern in lymph nodes (see Fig. 2.13 in Chap. 2).

Distinguishing Non-Hodgkin Lymphoma from Classical Hodgkin Lymphoma

Even once a diagnosis of malignant lymphoma is reached upon a frozen section, it may be quite difficult to distinguish, for example,

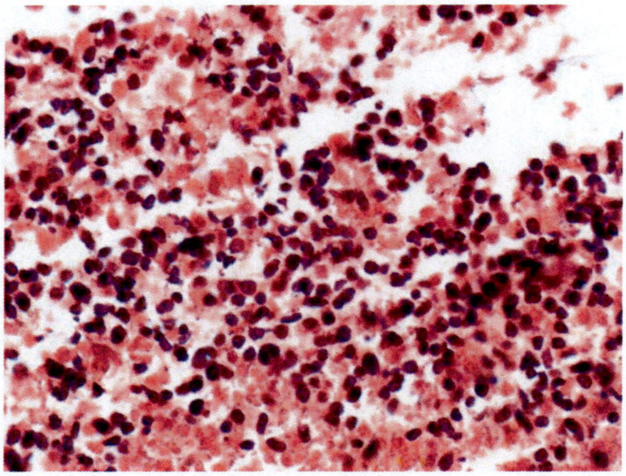

FIGURE 3.6 Frozen section of small, round, blue-cell tumor (Ewing sarcoma): Mimics malignant lymphoma with dyscohesive lymphoid-appearing cells.

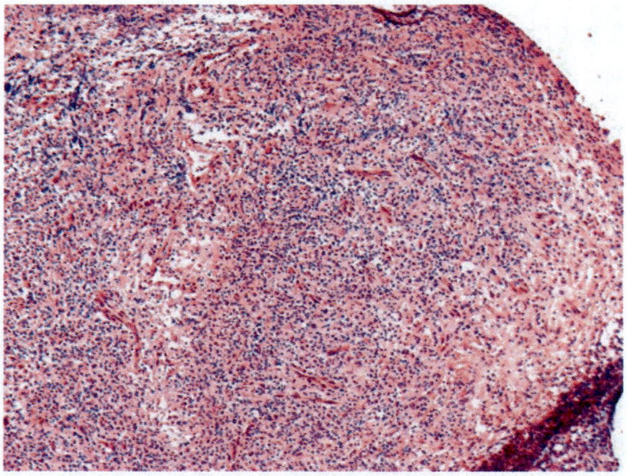

FIGURE 3.7 Frozen section of primary mediastinal B-cell lymphoma: Mimics nodular sclerosing classical Hodgkin lymphoma with nodule on low power surrounded by fibrosis and containing scattered large cells.

NS cHL from a primary mediastinal large B-cell lymphoma (PMBL), since this particular type of lymphoma may be associated with a marked degree of fibrosis and clusters of large abnormal mononucleated lymphoid cells (see Fig. 3.7). Of course, there are also those gray zone lymphomas with features intermediate between cHL and PMBL, which would not be able to be distinguished on a frozen section.

REFERENCES

1. Desciak EB, Maloney ME. Artifacts in frozen section preparation. Dermatol Surg. 2000;26:500–4.
2. Selak I. The Mohs laboratory. In: Mikhail G, editor. Mohs micrographic surgery. St. Louis: CV Mosby; 1991.
3. Cottell WI, Proper S. Mohs surgery, fresh-tissue technique: our technique with a review. J Dermatol Surg Oncol. 1982;8:576–8.
4. de Montpreville VT, Dulmet EM, Nashashibi N. Frozen section diagnosis and surgical biopsy of lymph nodes, tumors and pseudotumors of the mediastinum. Eur J Cardiothorac Surg. 1998;13(2):190–5.

Chapter 4
The Role of Fine Needle Aspiration (FNA) and FNA Biopsy in Evaluating Hematolymphoid Disorders and Malignancies in Lymph Nodes and Extranodal Tissues

INTRODUCTION

Fine needle aspirations (FNAs) and FNA biopsies (FNABs) may be indicated in evaluating lymph nodes and extranodal tissues for hematolymphoid disorders and malignancies instead of incisional or excisional biopsies performed in the operating room. The following situations are indications for these techniques and may be encountered in evaluating FNA and FNAB specimens from lymph nodes and extranodal sites in a "clinic" setting:

- Distinguishing reactive lymph nodes from malignant lymphoma [non-Hodgkin lymphoma (NHL) and classical HL (cHL)].
- Establishing an initial diagnosis of malignant lymphoma.
- Subclassification of malignant lymphoma in certain subtypes.
- Detecting recurrent lymphoma (NHL and cHL).
- Detecting composite lymphoma (NHL in background of cHL) and simultaneous hematolymphoid malignancies (NHL and acute myelogenous leukemia).
- Detecting hematopoietic malignancy (chloroma).
- Determining presence of a nonhematolymphoid malignancy.

C.H. Dunphy (ed.), *Frozen Section Library: Lymph Nodes*,
Frozen Section Library, 10, DOI 10.1007/978-1-4614-1253-3_4,

This chapter discusses the role of FNA and FNAB in evaluating hematolymphoid disorders in lymph nodes and extranodal tissues. However, before discussing these applications, the appropriate triaging procedures will be presented.

RECOMMENDED TRIAGE PROCEDURES FOR FNAS OF NON-LYMPH NODE SPECIMENS WITH A SUSPECTED HEMATOLYMPHOID MALIGNANCY AND FOR FNAS OF LYMPH NODE SPECIMENS

The first pass of the FNA that contains diagnostic material is retained for cytomorphological evaluation. The material from this first pass is expressed onto glass slides and smears are prepared, stained (one slide for Diff-Quick stain and one slide for Papanicalou stain), and reviewed. If by cytomorphological examination, there is obvious carcinoma, at least one additional pass is obtained for preparation of a cell block. If by cytomorphological examination, there is a suspicion of a hematolymphoid malignancy, a second pass is obtained and placed into Roswell Park Memorial Institute (RPMI) 1640 media (Cellgro media) for flow cytometric immunophenotyping (FCI). Additional passes may be obtained for additional material for FCI and/or other ancillary studies (to be discussed later in this chapter), and for FNA core biopsy (FNAB).

A cytospin preparation of the flow cytometry specimen is Wright-stained and reviewed to assess the quality of the specimen for FCI. It is recommended that the FNA flow cytometry specimen can be analyzed as soon as possible after collection; however, if necessary, it may be refrigerated overnight for processing the next morning.

These same guidelines apply to the triage of FNA specimens of lymph nodes in the following situations:

1. If there is a previous history of hematolymphoid malignancy (including but obviously not limited to cHL)
2. If suspicious for a hematolymphoid malignancy by cytomorphological examination
3. If there is a clinical suspicion of hematolymphoid malignancy
4. If the patient is older than 30 years of age

FCI should be routinely performed on specimens suspected of cHL, not only to exclude a B-cell NHL (which will most often reveal a monoclonal or aberrant B-cell population) or T-cell NHL (which may reveal an aberrant T-cell immunophenotype), but also to exclude the coexistence of cHL with a B-cell or T-cell NHL. In addition, FCI is recommended even when considering relapse of cHL, since a NHL may occur after therapy for cHL [1].

NORMAL LYMPH NODE FINDINGS BY FNA

FNAs of benign and reactive lymph nodes typically reveal a polymorphous mixture of predominantly small lymphocytes with intermixed larger forms and immunoblasts (see Figs. 4.1–4.3). Cytomorphological findings, such as numerous large "abnormal"

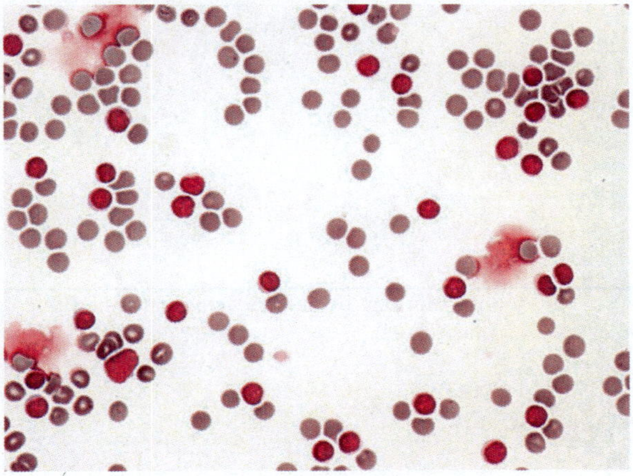

FIGURE 4.1 FNA cytomorphology of normal lymph node: Predominance of small lymphocytes with scattered larger forms.

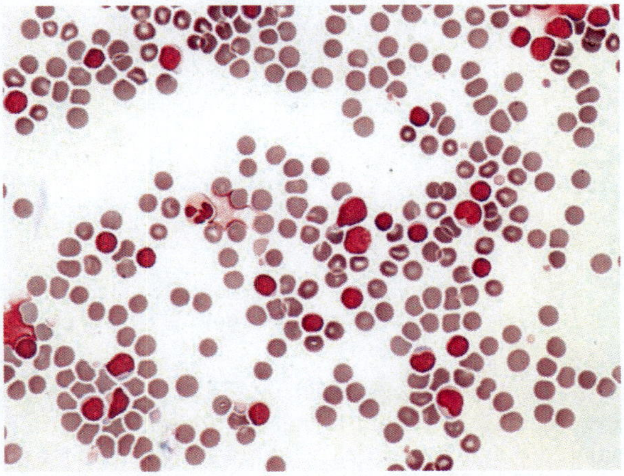

FIGURE 4.2 FNA cytomorphology of normal lymph node: Higher power of scattered larger forms and a polymorphonuclear cell.

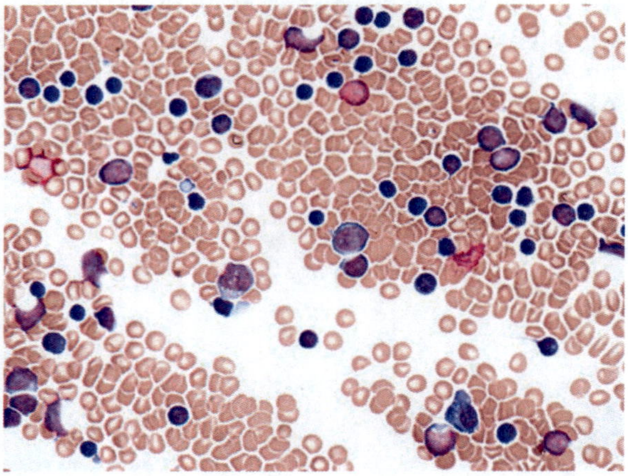

FIGURE 4.3 FNA cytomorphology of normal lymph node: Higher power of immunoblasts and a plasma cell.

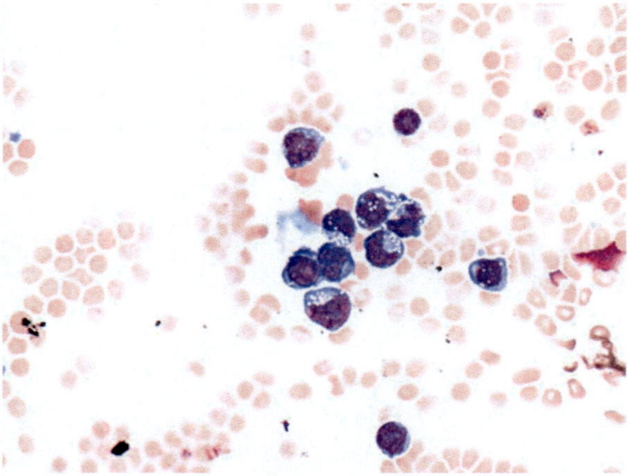

FIGURE 4.4 FNA cytomorphology of malignant non-Hodgkin lymphoma (large B-cell lymphoma): Numerous large abnormal lymphoid forms.

lymphoid forms (as demonstrated in Fig. 4.4), or a monotonous population of small lymphoid forms (as demonstrated in Fig. 4.5), are considered diagnostic or least highly suspicious of a malignant lymphoma. One should of course keep in mind that malignant lymphomas of both the non-Hodgkin and classical Hodgkin types may show a polymorphous population as well. Thus a polymorphous

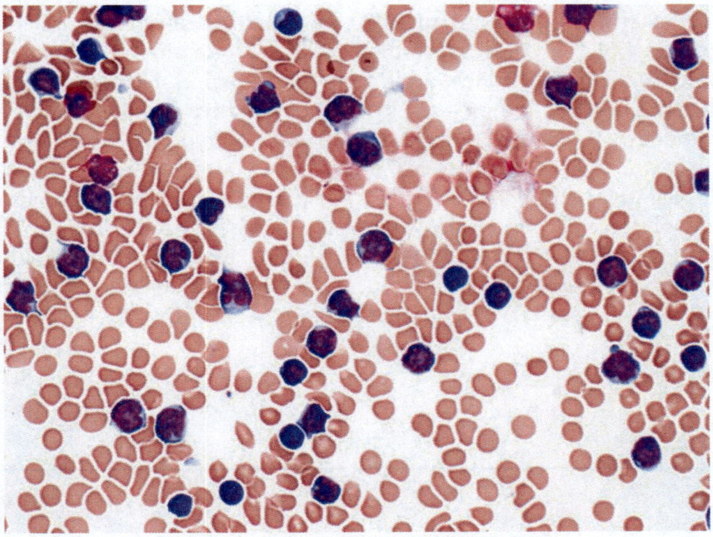

FIGURE 4.5 FNA cytomorphology of malignant non-Hodgkin lymphoma (small lymphocytic lymphoma): Very monotonous population of predominantly small cells.

population does not necessarily exclude involvement by a malignant lymphoma.

INITIAL DIAGNOSIS OF NON-HODGKIN LYMPHOMA

The cytomorphological features evaluated in specimens acquired by FNA combined with FCI have been reported to be successful in evaluating sites for lymphomatous involvement in 75–90% of cases [2, 3]. The detection of a monoclonal or aberrant B-cell population, significant lack of light chain expression on B cells, or an aberrant T-cell immunophenotype may aid in establishing a diagnosis of a lymphoid malignancy. B-cell clonality is defined as a kappa:lambda ratio of greater than 3:1 or less than 1:2. In addition, cytoplasmic expression of BCL2 may also be determined by flow cytometry. Malignant B cells demonstrate higher BCL2 expression than reactive B and T cells. BCL2 expression may also aid in establishing a correct diagnosis, especially in cases with no clear-cut monoclonality or no expression of light chains. The highest degree of BCL2 expression is characteristic of follicular lymphoma [4].

Significant lack of light chain expression on B cells may also aid in establishing a diagnosis of a B-cell lymphoma [5]. Shiyong

Li et al. reviewed more than 5,000 specimens received over a 6-year period; 1,561 of these specimens showed clonality. All cases, in which >25% B cells lacked sIg light-chain expression, represented lymphoma. In all these cases, the sIg-negative B-cell population was distinctly separate from the normal polytypic B lymphocytes. In 90% of these cases, forward angle light scatter indicated that the sIg-negative cells were larger than the background reactive T cells or the polytypic B cells. In those cases that had amplifiable DNA's, 71% of these showed a clonal IgH chain gene rearrangement. Further review of 500 reactive lymphoid populations showed no case with >17% sIg-B cells. Such a case example is seen in Fig. 4.6.

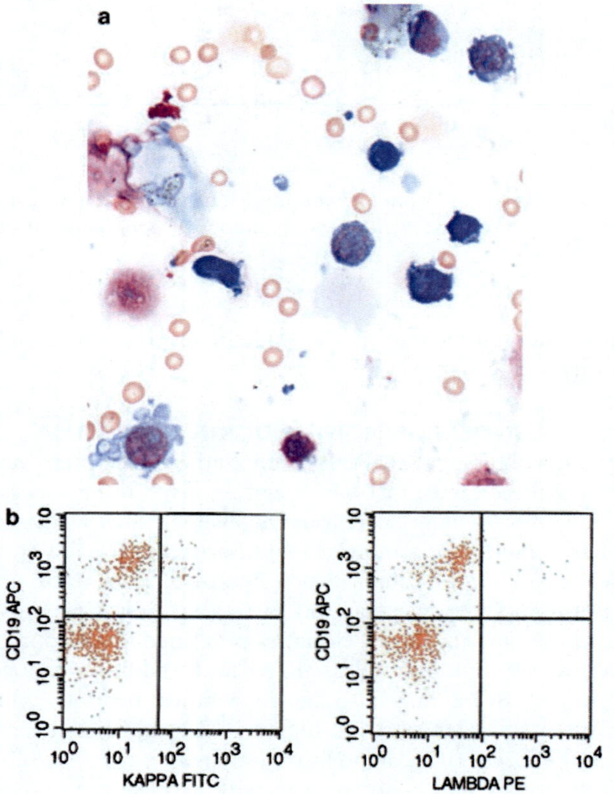

FIGURE 4.6 FNA of malignant non-Hodgkin lymphoma (large B-cell lymphoma) with lack of light chain expression: (**a**) Wright's stain of cytomorphology with large very abnormal lymphoid cells and (**b**) flow cytograms lacking any light chain expression (kappa and lambda – *upper right hand boxes*) expression with CD19 (*upper left hand boxes*).

Determination of an aberrant T-cell population by flow cytometry in a T-cell lymphoproliferative disorder may be more challenging than detecting an abnormal B-cell population in B-cell disorders. FCI findings in T-cell disorders are variable. There may be loss of a pan-T-cell antigen (i.e., CD2, CD3, CD5, and CD7). Most cases are CD4+; some are CD8+, CD4–/CD8–, or CD4+/CD8+. A T-cell lymphoma with an aberrant immunophenotype is demonstrated in Fig. 4.7.

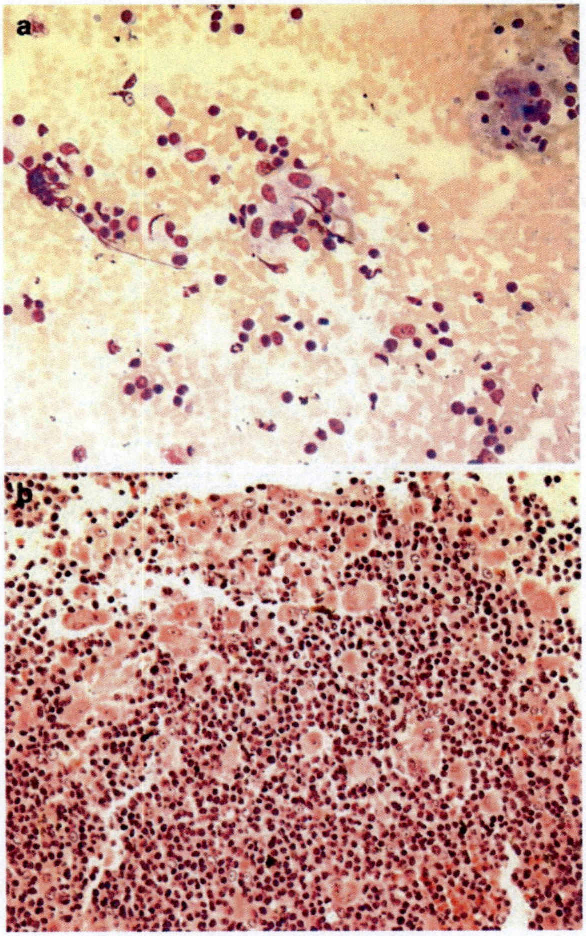

FIGURE 4.7 FNA of T-cell lymphoma: (**a**) Cytomorphology with a granulomata and scattered small lymphocytes in the background with lack of apparent cytologic atypia; (**b**) An H and E section of a subsequent biopsy reveals the presence of numerous epithelioid histiocytes in this T-cell lymphoma with irregular small lymphocytes in the background.

One must also remember that the finding of an aberrant T-cell immunophenotype by flow cytometry is helpful but must be correlated with histology.

The specificity of combining FNA cytomorphology and FCI in diagnosing NHLs has been repeatedly reported as 100%; however, the sensitivity is at best approximately 93% [2, 3, 6]. False "negatives" in the diagnosis of NHL may result from the following scenarios: tumoral necrosis or sclerosis, partial tissue involvement, T-cell NHL without an aberrant immunophenotype, or a T-cell- or lymphohistiocytic-rich diffuse large B-cell lymphoma (TCR-DLBCL or LHR-DLBCL) without detectable monoclonal B cells.

The ability and limitations of this combined approach in subclassifying (or subtyping) and grading of NHLs, in evaluating recurrent NHL, cHL, other hematolymphoid neoplasms, and non-hematolymphoid malignancies, in applying ancillary studies are further discussed in this chapter.

SUBCLASSIFICATION/SUBTYPING AND GRADING OF NON-HODGKIN LYMPHOMAS

Accurate subclassification of NHLs by FNA and FCI has been reported to be attainable in 71–79% of positive cases [6]. Subclassification is based on the abnormal, characteristic cytomorphology and immunophenotype, and thus is more likely attainable in certain subtypes of NHL. The small B-cell lymphomas/leukemias and their characteristic surface antigens are tabulated in Table 4.1. Subclassification is most easily, reliably attained in the following subtypes: small lymphocytic lymphoma/chronic lymphocytic leukemia (SLL/CLL and the large cell transformations), mantle cell lymphoma (MCL and the blastoid variants), Burkitt lymphoma (BL), plasma cell neoplasms, anaplastic large-cell lymphoma (ALCL), and precursor T-cell and B-cell lymphoblastic neoplasms. Such cases

TABLE 4.1 Small B-cell lymphomas/leukemias surface antigens.

Small B-cell lymphomas/leukemias surface antigens							
NHL	Sig	CD5	CD10	CD23	CD11c	CD103	CD25
CLL	dim	+	–	++/+	–	–	–
MCL	+	+	–/+	–/+	–	–	–
FL	+	–	+	+/–	–	–	–
LPL	+	–	–	–	–	–	–
MZL	+	–	–	–	+/–wk	–	–
SMZL	+	–	–	–	+/–wk	+/–wk	–
HCL	+	–	–/+	–	++	+	+

(i.e., CLL/SLL, MCL-classical and blastoid variants, BL-atypical and posttransplant cases, and ALCL) are demonstrated in Figs. 4.8–4.13, respectively. The accurate, reliable subclassification in these subtypes is due to the characteristic, uniform cytomorphology of the neoplastic cells and their unique, characteristic immunophenotypes. Ancillary studies to support these specific subtypes [i.e., t(11;14) in MCL, c-myc translocation in BL, t(2;5) in ALCL, etc.] may also be performed on FNA specimens, as discussed in the appropriate section below.

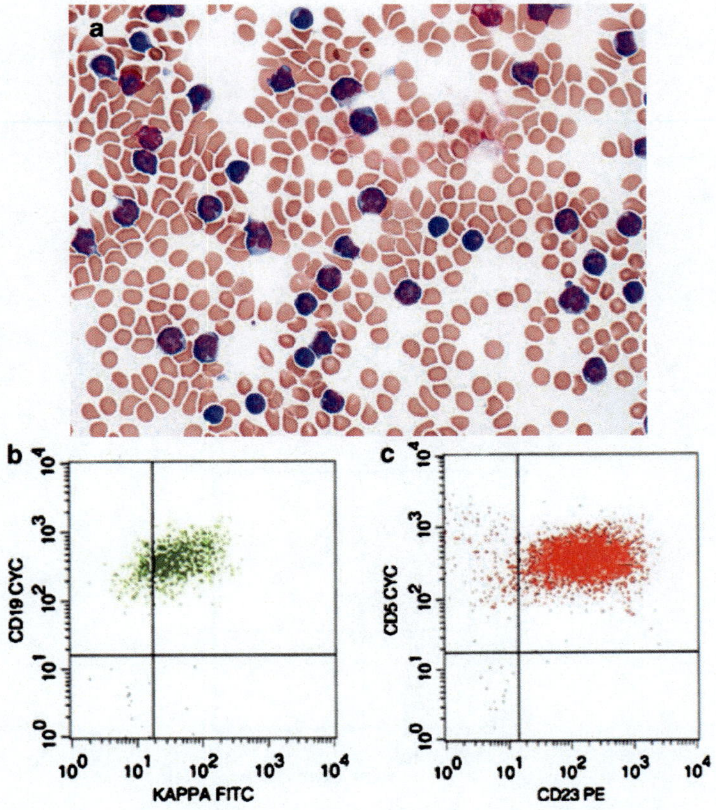

FIGURE 4.8 FNA of small lymphocytic lymphoma: (**a**) Cytomorphology with a monotonous population of numerous predominantly small lymphocytes; (**b**, **c**) reveal flow cytograms in this case demonstrating coexpression of CD19 and kappa [(**b**) *upper right hand box*] and CD5 and CD20 [(**c**) *upper right hand box*].

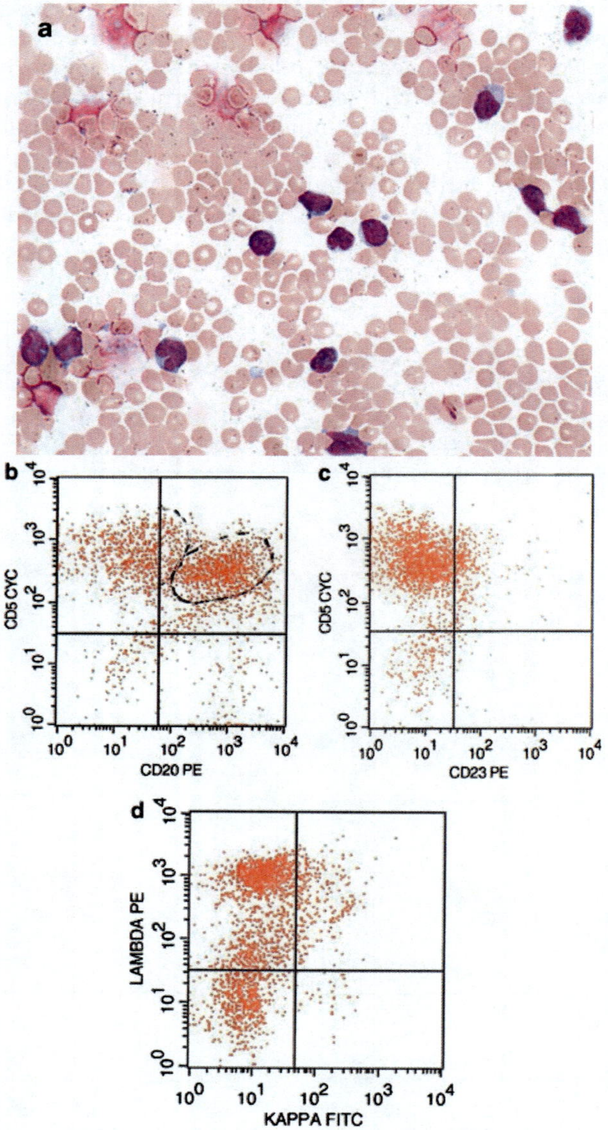

FIGURE 4.9 FNA of mantle cell lymphoma: (**a**) Cytomorphology composed of small slightly irregular lymphocytes; (**b–d**) reveal flow cytograms in this case demonstrating coexpression of CD5 and CD20 [(**b**) *upper right hand box*], lack of CD23 expression [(**c**) *upper right hand box*], and monoclonal expression of lambda [(**d**) *upper left hand box*].

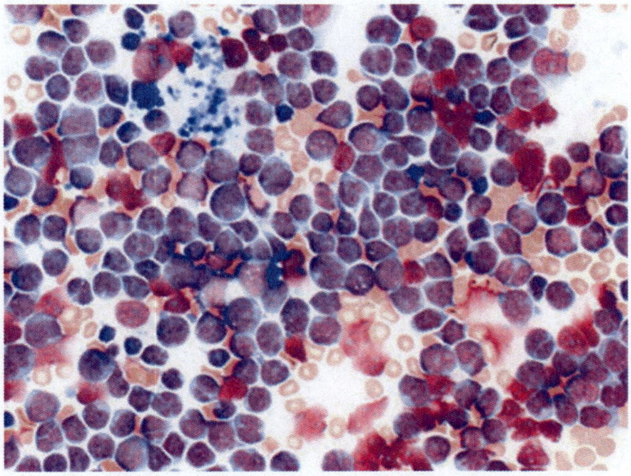

FIGURE 4.10 FNA cytomorphology of blastoid variant of mantle cell lymphoma: Numerous large blastic-appearing lymphoid forms.

Accurate, reliable subclassification of other NHL subtypes [i.e., marginal zone B-cell lymphoma, follicular lymphoma (FL), etc.] is not as easily attained by this approach. For example, the relative lack of architecture that is able to be evaluated in FNA specimens is a major limitation in subtyping FLs. FNA cytomorphology is not only limited in its ability to diagnose FL, but also to grade FLs. Merely counting the number of large cells may either overestimate or underestimate the number of large cells in the neoplastic population due to the presence of small and large reactive lymphocytes, the crushing of fragile large neoplastic cells, and the morphologic variability of transformed lymphoid cells. More recently, evaluation of grading FLs using ThinPrep slides has shown a significant upward trend in the number of centroblasts as the grades of FL increased. On the other hand, analysis of cell size by flow cytometry is not as reliable as using ThinPrep slides to distinguish grades of FL, especially Grade 2 versus Grade 3 [7]. In fact, although flow cytometry has the ability to analyze a large number of cells than by cytomorphological counting of large cells, it has the propensity to underestimate the number of large cells due to their lower viability.

Even with ThinPrep slides, the grading of FLs by FNA combined with FCI, continues to be limited, due to the simple, undeniable fact that grading of FL by the WHO classification is based on the average number of centroblasts or large "transformed" cells in ten neoplastic follicles at 40× high-power field (hpf) examination

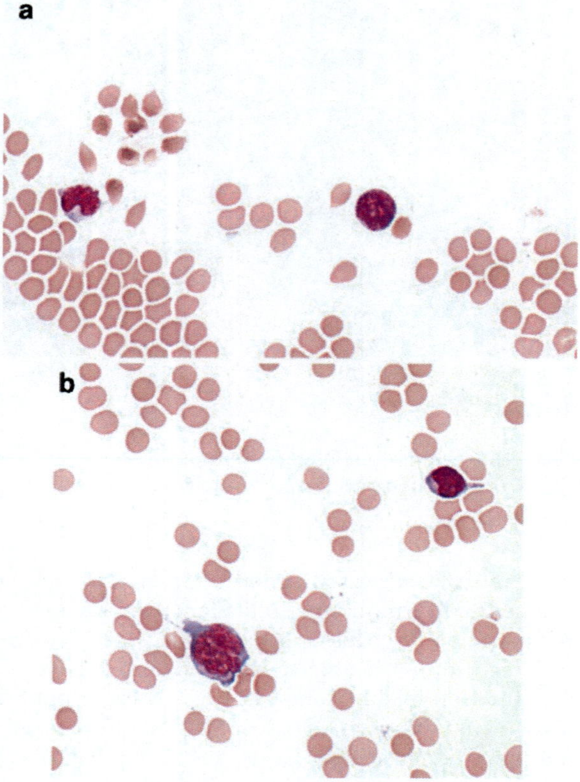

FIGURE 4.11 FNA cytomorphology of atypical Burkitt lymphoma (BL): (**a**) Single abnormal intermediate-sized abnormal lymphoid cell typical of BL; (**b**) larger abnormal lymphoid cell as may be seen in the atypical variant of BL. This case revealed a CD10+ monoclonal B-cell population by flow cytometry, as is typical of BL and a c-myc gene rearrangement.

(i.e., Grade 1: 0–5 centroblasts/hpf; Grade 2: 6–15 centroblasts/hpf; and Grade 3: greater than 15 centroblasts/hpf) [8]. It is important to accurately, reliably grade FLs, since grading of FL, in combination with clinical factors, is what ultimately effects prognosis and therapeutic decisions in these patients.

Although grading of FL is not entirely reliable by combining FNA and FCI, the evaluation of cell size by combining FCI and FNA cytomorphology does have diagnostic and clinical applications. The combined approach has been demonstrated to be useful in reliably diagnosing large-cell lymphoma/transformation, if

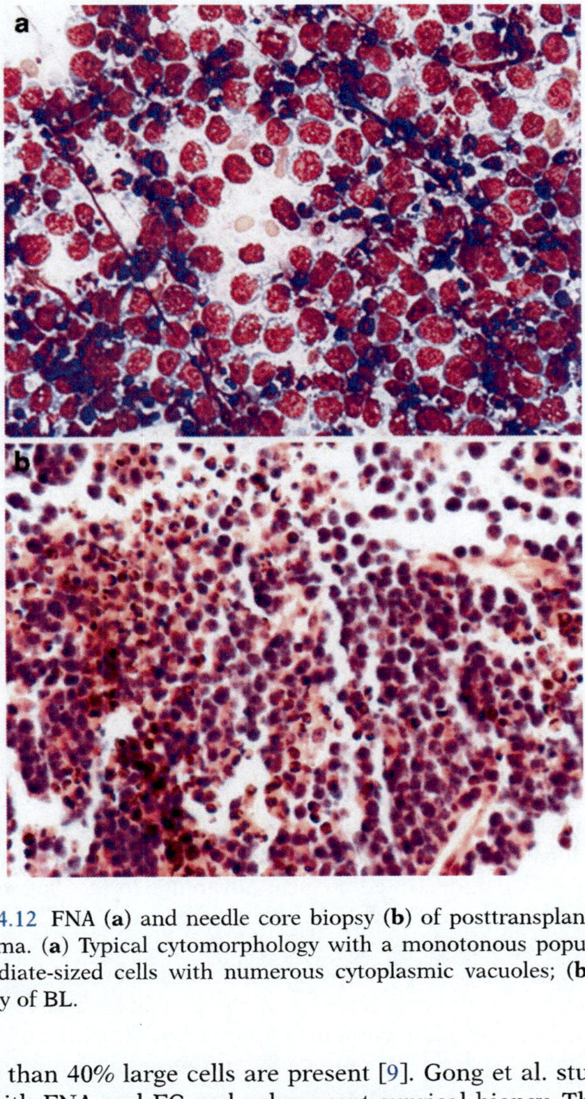

FIGURE 4.12 FNA (**a**) and needle core biopsy (**b**) of posttransplant Burkitt lymphoma. (**a**) Typical cytomorphology with a monotonous population of intermediate-sized cells with numerous cytoplasmic vacuoles; (**b**) typical histology of BL.

greater than 40% large cells are present [9]. Gong et al. studied 48 NHL with FNA and FC and subsequent surgical biopsy. The study evaluated 200 lymphocytes in each specimen by cytomorphology. All small-cell NHLs showed <40% large cells by CM and FC. Diffuse large B-cell lymphoma (DLBCL) demonstrated >40% large cells by FC (all cases) and by CM (67% cases). This study detected variable numbers of large cells in grade 3 FL, low-grade lymphoma with partial large cell transformation, and DLBCL containing fewer than

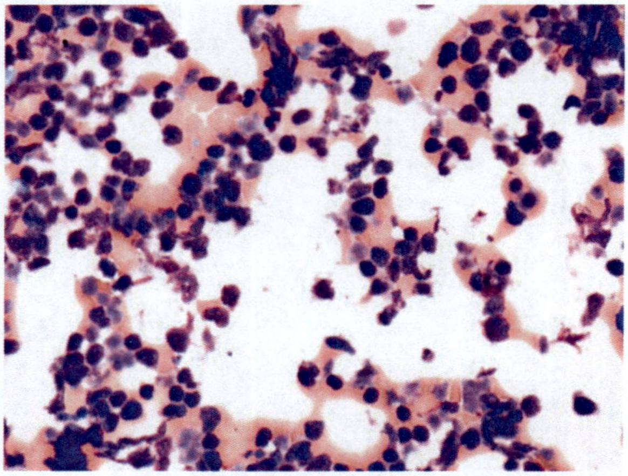

FIGURE 4.13 FNA cytomorphology of CD30+ anaplastic large-cell lymphoma: Numerous abnormal lymphoid forms with occasional "wreath" and "doughnut" forms.

10% neoplastic cells. In conclusion, large-cell lymphoma and large-cell transformation may be reliably diagnosed if >40% large cells are encountered.

DIAGNOSTIC IMPACT OF CORE NEEDLE BIOPSY

Since there are limitations, as discussed previously, regarding the subclassification and grading of NHLs, the question arises as to the contribution (or diagnostic impact) of obtaining and evaluating a core needle biopsy. A rather large retrospective study of 74 FNA specimens by Gong et al. demonstrated no clear advantage in the diagnosis and classification of small B-cell NHLs by adding a core needle biopsy to FNA and FCI [10]. However, a core needle biopsy was shown to be more useful in diagnosing DLBCL: 37% of cases were able to be diagnosed when a core needle biopsy was also available versus only 25% by FNA and FCI. Thus, a combined approach reduces the number of insufficient cases and is recommended in routine FNA practice.

VALUE OF FLUORESCENT IN SITU HYBRIDIZATION AND POLYMERASE CHAIN REACTION ANALYSIS IN THE DIAGNOSIS OF NHL BY FNA

Although FNA specimens may not provide analyzable metaphases for conventional cytogenetic studies, they are well-suited to directed fluorescent in situ hybridization (FISH) analyses [i.e., t(14;18) (or BCL2) for a definitive diagnosis of FL; t(11;14) (or BCL1) for MCL; a c-myc rearrangement for BL; t(2;5) for ALCL, etc.]. In addition, with proper handling and management of specimens, FNA can routinely provide samples adequate for molecular genetic studies. In a study by Safley et al., of 30 FNAs with suspected hematolymphoid malignancies (20 diagnosed as B-cell NHL), the additional application of FISH and polymerase chain reaction (PCR) analyses definitely subclassified 12 of the B-cell NHLs [i.e., 4 FLs by BCL2 FISH or PCR, 2 MCLs by BCL1 FISH, and 50% of SLL (2/4) and DLBCL (4/8)] [11]. With proper handling and management of specimens, FNA can routinely provide samples adequate for molecular genetic studies.

EVALUATION OF RECURRENT NHL

The combined approach of FNA and FCI is also useful in evaluating tissues for recurrent involvement by a NHL. In a large, retrospective study of this approach in the evaluation of recurrent NHL, Chernoff et al. demonstrated that 90% of recurrent cases were amenable to FNA [12]. In addition, cytomorphology identified large cell transformation, as defined previously. A later study further pointed out that the findings of necrosis and numerous polymorphonuclear cells in this setting demand a tissue biopsy, since all of these cases represented recurrent NHL upon follow-up biopsy [13].

LIMITATIONS OF FNA COMBINED WITH FCI AND FNAB IN THE EVALUATION OF PRIMARY AND RECURRENT LYMPHOMATOUS INVOLVEMENT

As discussed previously, false negatives may result from the following situations: sampling issues, tumoral necrosis or sclerosis, partial tissue involvement, a T-cell NHL without an aberrant immunophenotype, or a TCR-DLBCL or LHR-DLBCL without detectable monoclonal B cells (see Table 4.2). In addition, there are limitations in definitive subclassification and grading of NHLs. Architecture (i.e., follicular

TABLE 4.2 Evaluation of lymphomatous involvement by FNA and FCI: Limitations and situations requiring large incisional or excisional biopsy.

Situations with "false-negative" results	Situations requiring large incisional/ excisional bx
Tumoral necrosis	Grading of FL
Tumoral sclerosis	"Indeterminant" or suspicious results
T-NHL without an aberrant IP	Necrosis: "initial" evaluation for NHL
B-NHL with T-cell- or LH-rich background	Necrosis and numerous PMNs: evaluation for recurrent NHL

(continued)

TABLE 4.2 (continued)

Situations with "false-negative" results	Situations requiring large incisional/ excisional bx
Classical HL	Fewer than 10% neoplastic cells detected by FCI
Sampling issues/discordance	Predominance of small cells by cyto-morphology or by FCI
	With signs of clinical transformation
	Evaluating for an initial dx of cHL

FNA fine needle aspiration, *FCI* flow cytometric immunophenotyping, *NHL* non-Hodgkin lymphoma, *IP* immunophenotype, *FL* follicular lymphoma, *LH* lymphohistiocytic, *PMNs* polymorphonuclear cells, *cHL* classical Hodgkin lymphoma, *dx* diagnosis, *bx* biopsy

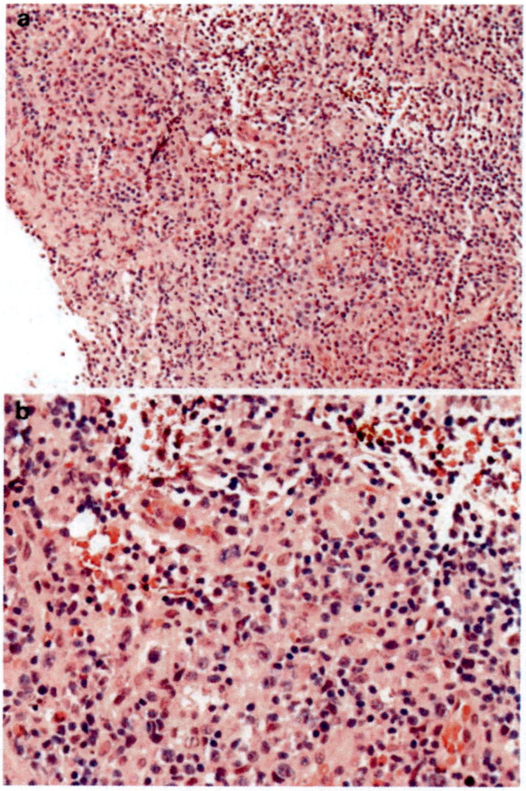

FIGURE 4.14 Histology of needle biopsy of EBV-infected lymph node: Mimics large-cell lymphoma [(**a**) low power; (**b**) higher power].

vs. diffuse) is not able to be determined by FNA cytomorphology. One should also exercise extreme caution in determining such architecture based only on a thin core needle biopsy. Core needle biopsies of lymph node specimens may be overinterpreted as large-cell lymphoma (i.e., if the needle happens to pass through a large, reactive germinal center, through an EBV-infected lymph node with marked reactive changes, etc.) See a case example in Fig. 4.14. As also previously mentioned, recurrent NHL is often associated with no definitive cytomorphological evidence of lymphomatous involvement, but rather a background of necrosis and numerous polymorphonuclear cells, which demands a tissue biopsy. An initial diagnosis of cHL also requires a tissue biopsy, even if able to be diagnosed by immunophenotyping of the FNA or core needle biopsy, to determine the subtype, which is based on architectural patterns.

CLASSICAL HODGKIN LYMPHOMA: PRIMARY DIAGNOSIS AND RECURRENCE

The combined approach of using FNA and FCI is useful in evaluating tissues for involvement by cHL, either as an initial diagnosis or as a recurrence. This combined approach primarily allows for the exclusion of the presence of a NHL, either alone or as a composite lymphoma. In addition, FCI is recommended even when considering relapse of cHL, since a NHL may occur after therapy for cHL.

COMPOSITE LYMPHOMA AND SIMULTANEOUS HEMATOLYMPHOID MALIGNANCIES

FNA and FCI may be useful not only in identifying a composite lymphoma of cHL and NHL, but also in detecting a composite lymphoma of a B-cell NHL and a T-cell NHL. This combined approach also allows for the simultaneous presence of two hematolymphoid malignancies, for example, a coexistent NHL and acute myelogenous leukemia (AML).

SITUATIONS REQUIRING BIOPSY, BASED ON FNA AND FCI RESULTS (TABLE 4.2)

The following situations require biopsy, based on FNA and FCI results, when evaluating lymphomatous involvement:

1. NHL of follicle center cell origin with a mixed cellular composition (to determine exact cellular composition, pattern of involvement, and to exclude a component of DLBCL).
2. Indeterminate results.
3. The presence of necrosis in the evaluation of an initial diagnosis of lymphoma.

4. The presence of necrosis and polymorphonuclear cells in the evaluation of recurrent NHL.
5. Fewer than 10% neoplastic cells detected by FCI (to determine the significance of the flow cytometric finding).
6. Finding a predominance of small cells by cytomorphology or by FCI with clinical signs of transformation (to exclude a large cell component).
7. Evaluating for an initial diagnosis of cHL (for exact subtyping and to evaluate a composite lymphoma).

DETECTING HEMATOPOIETIC MALIGNANCY [I.E., GRANULOCYTIC SARCOMA (CHLOROMA), MONOCYTIC SARCOMA, ERYTHROID SARCOMA]

Involvement of tissue specimens by granulocytic, monocytic, or erythroid sarcomas should be considered in the differential diagnosis of NHLs of B-cell and T-cell origin, respectively, due to the similarities of the blasts to large-cell lymphoma, the presence of lymphoglandular bodies, and the rarity of Auer rods and eosinophilic myelocytes. The combined approach of FNA and FCI may be quite useful in these differential diagnoses. The neoplastic cells of granulocytic sarcomas are typically distinguished by their variable expression of myelomonocytic markers (i.e., CD11b, CD13, CD15, CD33, CD64, etc.), as AMLs. They most commonly express CD13 and CD33 with variable expression of CD34 [14], and are negative for CD10, CD20, and surface light chains. They may aberrantly express CD19 or T-cell antigens, as in AMLs. One should keep in mind that AMLs associated with t(8;21) are also associated with PAX-5 nuclear positivity by immunohistochemical staining. The neoplastic cells of monocytic sarcomas, like AMLs with monocytic differentiation, typically show variable expressions of CD13, CD11b, CD14, CD15, and CD33 with generally strong expressions of CD64 and HLA-Dr. They are typically CD34-negative. The neoplastic cells of erythroid sarcoma may be distinguished from NHLs by their lack of B-cell and T-cell antigens and expression of glycophorin A (see Fig. 4.15). Also, one should keep in mind that ALCL may express myelomonocytic markers, but the expression of CD30, variable expression of T-cell markers, and lack of immature markers should distinguish ALCL from granulocytic sarcoma.

DETERMINING PRESENCE OF METASTATIC NONHEMATOLYMPHOID MALIGNANCY

A combined FCA and FCI approach may also be useful in determining the presence of a nonhematolymphoid malignancy by detecting cohesive malignant cells by cytomorphology and the presence of

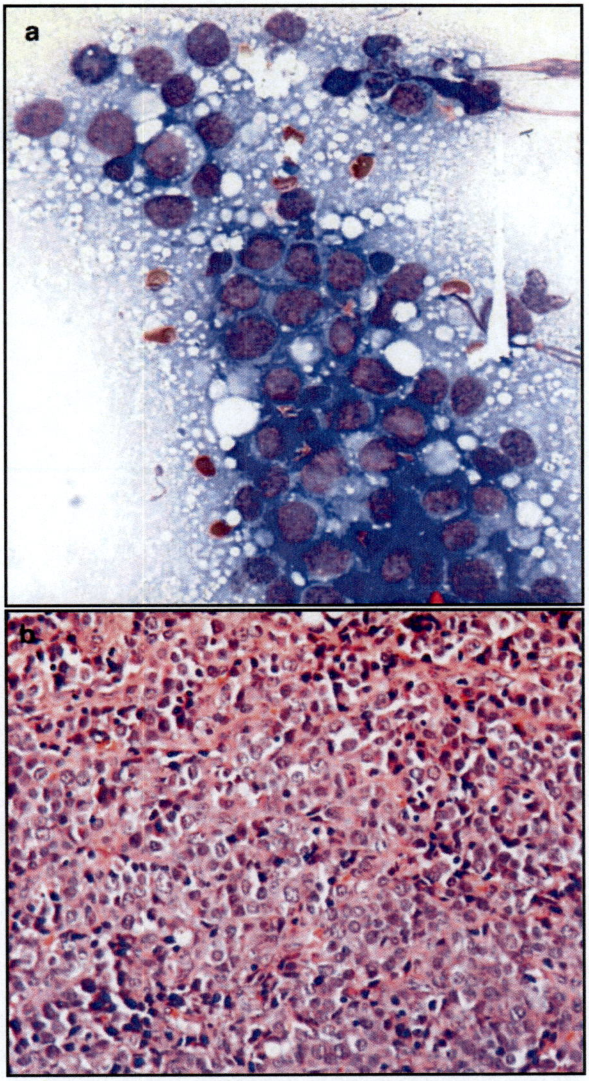

FIGURE 4.15 Touch preparation cytomorphology and histology of erythroid sarcoma: (**a**) TP cytomorphology with blasts associated with light blue cytoplasm-containing vacuoles; (**b**) histology reveals dyscohesive large cells resembling large-cell lymphoma and (**c**) HbA stain confirms the erythroid origin of the malignant cells.

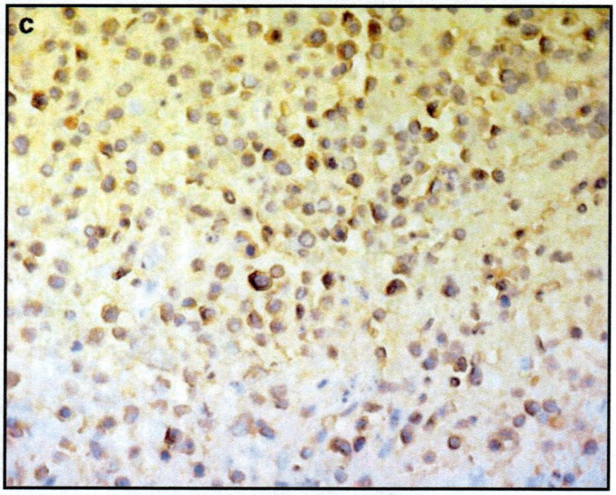

FIGURE 4.15 (continued)

large, CD45-negative cells by FCI. However, one should keep in mind that some cases of large-cell lymphoma may also be CD45-negative, particularly ALCL, so correlation with the cytomorphology and additional markers are necessary to exclude a diagnosis of lymphoma in such cases.

REFERENCES

1. Borchmann P, Behringer K, Josting A, et al. Secondary malignancies after successful primary treatment of malignant Hodgkin's lymphoma. Pathologe. 2006;27:47–52.
2. Dong HY, Harris NL, Preffer FI, Pitman MB. Fine-needle aspiration biopsy in the diagnosis and classification of primary and recurrent lymphoma: a retrospective analysis of the utility of cytomorphology and flow cytometry. Mod Pathol. 2001;14(5):472–81.
3. Dey P, Amir T, Jassar AA, et al. Combined applications of fine needle aspiration cytology and flow cytometric immunophenotyping for diagnosis and classification of non Hodgkin lymphoma. CytoJournal. 2006;3:24.
4. Cook JR, Craig FE, Swerdlow SH. Bcl-2 expression by multicolor flow cytometric analysis assists in the diagnosis of follicular lymphoma in lymph node and bone marrow. Am J Clin Pathol. 2003;119:145–51.
5. Li S, Eshleman JR, Borowitz MJ. Lack of surface immunoglobulin light chain expression by flow cytometric immunophenotyping can help diagnose peripheral B-cell lymphoma. Am J Clin Pathol. 2002;118: 229–34.

6. Zeppa P, Marion G, Troncone G, et al. Fine-needle cytology and flow cytometry immunophenotyping and subclassification of non-Hodgkin lymphoma: a critical review of 307 cases with technical suggestions. Cancer Cytopathol. 2004;102:55–65.
7. Brandao GDA, Rose R, Mckenzie S, Maslak P, Lin O. Grading follicular lymphomas in fine-needle aspiration biopsies: the role of ThinPrep slides and flow cytometry. Cancer Cytopathol. 2006;108:319–23.
8. Jaffe ES, Harris NL, Stein H, Vardiman JW. World Health Organization classification of tumours. Pathology and genetics of tumours of haematopoietic and lymphoid tissues. Lyon: IARC Press; 2001.
9. Gong JZ, Williams Jr DC, Liu K, Jones C. Fine-needle aspiration in non-Hodgkin lymphoma: evaluation of cell size by cytomorphology and flow cytometry. Am J Clin Pathol. 2002;117(6):880–8.
10. Gong JZ, Snyder MJ, Lagoo AS, et al. Diagnostic impact of core-needle biopsy on fine-needle aspiration of non-Hodgkin lymphoma. Diagn Cytopathol. 2004;31(1):23–30.
11. Safley AM, Buckley PJ, Creager AJ, et al. The value of fluorescence in situ hybridization and polymerase chain reaction in the diagnosis of B-cell non-Hodgkin lymphoma by fine-needle aspiration. Arch Pathol Lab Med. 2004;128(12):1395–403.
12. Chernoff WG, Lampe HB, Cramer H, Banerjee D. The potential clinical impact of the fine needle aspiration/flow cytometric diagnosis of malignant lymphoma. J Otolaryngol. 1992;21 Suppl 1:1–15.
13. Dunphy CH, Ramos R. Combining fine-needle aspiration and flow cytometric immunophenotyping in evaluation of nodal and extranodal sites for possible lymphoma: a retrospective review. Diagn Cytopathol. 1997;16:200–6.
14. Suh YK, Shin HJC. Fine-needle aspiration biopsy of granulocytic sarcoma: a clinicopathologic study of 27 cases. Cancer Cytopathol. 2000;90:364–72.

Chapter 5

Applications of Frozen Section to Intraoperative Consultations of Metastatic Malignancies Involving Lymph Nodes

INTRODUCTION

Intraoperative assessment of lymph nodes can be a powerful tool capable of guiding clinical care by revealing diagnoses and clarifying staging at the time of a patient's operative procedure. When used appropriately, it has the potential to spare a patient from the risk and cost of unnecessary procedures and, in some instances, to identify the need for more extensive surgical intervention without the need for an additional episode of general anesthesia. However, there are important limitations to intraoperative evaluation of lymph nodes in such situations, of which both clinicians and pathologists must be cognizant.

Before a lymph node is analyzed by frozen section (and even more proximately, before it is sent to the frozen section room), one should consider a question fundamental to all intraoperative consultations: Will information from the frozen section diagnosis alter the present operative management in a meaningful way [1]? If not, intraoperative frozen section should not be performed, as it increases costs, introduces irreversible artifacts of freezing tissue that may impede pathologic evaluation, and decreases the tissue available for permanent section analysis [2]. In fact, as discussed below, many theoretical or historical applications of lymph node

C.H. Dunphy (ed.), *Frozen Section Library: Lymph Nodes*,
Frozen Section Library, 10, DOI 10.1007/978-1-4614-1253-3_5,

frozen sections have been abandoned precisely because they failed to meet this fundamental criterion.

Another factor that potentially limits the utility of frozen section analysis of lymph nodes is the quality of the histologic sections, which depends on numerous variables, including (but not limited to) tissue composition (i.e., it may be difficult to obtain good histologic sections of fatty-replaced lymph nodes) and the skill of the individual preparing the sections. When compared with histologic sections prepared from formalin-fixed, paraffin-embedded tissue, the shortcomings of sections prepared from frozen tissue include greater irregularity in staining with hematoxylin and eosin, increased tissue tears, cracks and folds, air-drying artifacts, etc. [3]. These technical aspects are discussed later in this chapter. The net result is that the pathologist is left to interpret inferior quality sections, with tissue architecture that may be obscured and disrupted by processing. In some cases, such as sentinel lymph node (SLN) biopsy in melanoma, the quality of the diagnosis has not been adequate to justify intraoperative consultation, as discussed below.

Despite these limitations, there are a number of clinical scenarios in which lymph node frozen sections may alter operative management in real time. Most of these cases fall into one of the two general categories. Probably, the most well known and best studied are SLN biopsies, in which the presence of metastatic disease is an indication for a more extensive lymph node dissection. In other instances, the presence of nodal metastases classifies disease as an inoperable, in which case lymph node frozen sections are performed to determine whether the primary tumor can be treated surgically. In these cases, the finding of metastatic disease in a lymph node at the time of the procedure can inform the surgeon that the procedure should be aborted or altered, thus sparing the patient a more extensive operation. More specific indications for lymph node frozen sections in patients with known or suspected malignancy are considered below.

SENTINEL LYMPH NODE MAPPING AND BIOPSY

The observation that labeled markers could be used to demonstrate the path of cancer spread through the lymphatic system dates back to the mid-1950s [4]. The term "sentinel lymph node" was coined just a few years later to describe the first draining node in patients with parotid cancer [5]. However, it was not until the early 1990s that this principle was first applied to intraoperative staging. Morton and colleagues performed the seminal studies demonstrating the clinical utility of SLN mapping and biopsying in a series of patients with melanoma [6]. In the nearly two decades since, SLN mapping with gamma-labeled probes and vital dyes has been applied to a number of malignancies with varying clinical utility.

SLN Biopsy and Breast Cancer

Currently, the most widely practiced use of SLN biopsy is in staging of T1 and T2 breast carcinomas. The utility of SLN mapping in breast cancer was first demonstrated in the mid-1990s. The initial work did not focus on intraoperative consultation, but rather demonstrated the value of sentinel node mapping for the identification of metastatic tissue in permanent sections [7, 8]. Giuliano and colleagues showed a 13% increase in node-positive breast cancers with SLN mapping, when compared with more traditional axillary dissections. This increased detection likely reflects the value of increased diagnostic attention to the draining node, allowing the pathologist to identify micrometastatic disease that would have been missed by traditional methods. In their seminal work, Giuliano et al. [7] examined six to eight levels on each sentinel node, compared with one to two levels on nodes from standard dissections, and also performed cytokeratin immunostaining on all labeled specimens. These protocols allowed the identification of smaller foci of disease in permanent sections and set the stage for intraoperative assessment of lymph nodes.

In Giuliano's initial studies, intraoperative consultation was utilized only for confirmation that labeled tissue was in fact lymphatic. Today, however, intraoperative consultation is used as a means of determining if a full axillary lymph node dissection is necessary in a patient with clinically node negative, T1 or T2 breast cancer. Prior to the development of SLN procedures, operative management of clinically node-negative breast cancer included level I and II axillary node dissections [9]. However, 70% of clinically node-negative patients also showed no histologic evidence of lymph node disease, meaning that a high proportion of women underwent full axillary node dissection unnecessarily. As a result, these women faced increased risk of infection and lymphedema, despite receiving no clinical benefit from the procedure [9]. SLN biopsy allows clinicians to overcome this problem by identifying the subset of patients with nodal disease, in whom the potential benefits of complete axillary lymphadenectomy might outweigh the morbidity associated with the more extensive procedure.

During the last decade, the accuracy of frozen section evaluation of SLNs in women with breast cancer has been considered in a large number of studies. The reported sensitivity and specificity of intraoperative assessments are determined in comparison to the accepted standard assessment on histologic evaluation of permanent section slides. The possible errors include false-negative frozen section results (tumor is detected on permanent section but not at the time of intraoperative assessment) and false-positive frozen section results (tumor is incorrectly identified at the time of intraoperative assessment but not identified on permanent section slides or review of frozen sections), both of which may have important clinical implications for the patient. False-negative results with metastatic disease recognized on permanent sections may necessitate a second procedure, with associated costs and risks of anesthesia. False-positive results are potentially more damaging, as the patient is exposed to the unnecessary morbidity of a complete axillary lymph node dissection. Fortunately, false-positive results are rare, and reported specificities have been 100% in most recent studies [10–15]. Sensitivity of frozen section evaluation of SLN has been much more variable, with results ranging from 52 to 95% depending on the study, with at least some of this variability due to differences in methodology [10–16]. The fact that reported sensitivities of SLN frozen sections in women with breast cancer have tended to increase over time suggests that intraoperative assessments are improving as techniques are refined and as pathologists become more experienced evaluating these specimens [15].

Despite improvements over time, the relatively low sensitivity (with a negative intraoperative assessment) means that a significant number of patients require subsequent completion of the axillary node dissection, when metastatic carcinoma is identified on permanent sections. These "false negatives" may occur for a variety of reasons. Some false negatives arise as a result of the small volume of disease (micrometastatic foci, by definition measuring less than 2 mm in diameter), which may be present but not recognized, or simply not be present in the sections of lymph node assessed intraoperatively. There has been much debate about the significance of these smaller disease foci, and whether they share the same prognostic significance as larger metastatic foci. Supporting the clinical importance of micrometastases, even areas of tumor less than 2 mm in SLNs may be associated with macrometastatic disease in other axillary nodes [17]. However, studies have not consistently shown the differences in overall and disease-free survival between patients with pN1mi disease (micrometastases between 0.2 and 2 mm) and patients with negative lymph nodes (pN0) [18–20]. As a result, it remains unclear whether all false-negative SLN biopsies need to be followed by completion axillary dissection. Furthermore, these results call into question the use of intraoperative detection of a micrometastasis as grounds for a completion axillary lymph node dissection.

Intraoperative evaluation of SLNs is not limited to the evaluation of frozen sections. Imprint cytology (IC) may also be used to identify nodal metastases. This approach has the advantage of avoiding freezing artifacts and decreasing tissue loss. However, nodal architecture is not retained with IC, which can make accurate diagnosis more difficult. In addition, a fibrotic stromal response may lead to few tumor cells being present on the imprint slides even in the setting of more extensive disease. Moreover, evaluation of imprints requires a degree of cytopathology expertise that may not be available at all centers. In the largest prospective, head-to-head comparison to date, the sensitivity and specificity of frozen section evaluation and IC were similar, with both approaches proving inadequate for the detection of micrometastases [21]. However, in a meta-analysis combining data from three earlier studies, frozen section evaluation was found to be more sensitive than IC (76 versus 62%) with no difference in specificity (99%) [22].

Some have recommended that intraoperative frozen sectioning can be supplemented with rapid immunohistochemistry to increase the sensitivity for identifying micrometastatic disease. Cytokeratin staining can be employed in the assessment of SLN in breast adenocarcinoma [23]. In one study of 100 patients with invasive mammary carcinoma, frozen section evaluation with rapid cytokeratin immunostaining was not only superior to both frozen section evaluation and IC but also approached the sensitivity of permanent section evaluation [24]. However, published studies do not consistently show clinical benefit of intraoperative IHC, and the increased processing time has several undesirable consequences, including increased cost and increased duration of anesthesia. Finally, benign epithelial inclusions may occasionally be found in lymph nodes (see Chap. 6), and the increased sensitivity of cytokeratin immunohistochemical studies may lead to additional false-positive results.

Despite the impressive body of evidence supporting the accuracy of intraoperative assessment of lymph nodes in patients with breast cancer, the technique has an uncertain future. Several studies have called into question whether axillary dissection should be performed, even in women with known positive SLNs [25, 26]. Most notably, the American College of Surgeons Oncology Group Z011 trial randomized women with positive sentinel nodes to be treated with or without completion axillary dissection. Despite some limitations (including a smaller than expected sample size and relatively short duration of follow-up), there was no difference in overall survival between study groups, suggesting that axillary lymph nodedissection may not always be necessary even in the setting of a positive SLN [26]. If this is indeed proven to be the case, then intraoperative assessment of SLNs will have an extremely limited scope of utility. These results may result in significant changes in management of patients with breast cancer. Today, most women with T1 or T2 breast cancer typically receive systemic adjuvant chemotherapy and tangential field irradiation of level I and II axillary nodes, regardless of SLN status. The Z011 study suggests that these therapies are effective for treatment of nodal spread and that axillary lymph node dissection may not provide additional benefit. If subsequent studies support this initial assessment, intraoperative evaluation of SLNs for patients with breast cancer should be restricted to a carefully selected subset of breast cancer patients, or perhaps even eliminated altogether.

Other Applications of SLN Frozen Section Evaluation

SLN biopsy has been used to assess a number of other tumors, primarily in research settings. As mentioned above, the seminal use of the approach was actually in intraoperative assessment of lymph nodes in patients with melanoma. However, in several large series, the sensitivity of SLN frozen section in patients with melanoma has ranged from 29 to 82%. This level of sensitivity is much lower than what has been reported for breast carcinoma. As a result, the frozen section analysis of lymph nodes in patients with melanoma is no longer recommended [27].

One other, albeit controversial, application of SLN biopsy has been in cases of gastric adenocarcinoma. In Japan, SLN biopsy is routinely used to evaluate the need for D2 lymphadenectomy in patients with T1 gastric cancers [28]. However, studies in Western nations have failed to demonstrate a clear survival advantage to D2 dissection, and as a result, SLN biopsies are not performed in patients with gastric cancer in the USA or Europe [29].

ASSESSING UTILITY OF OPERATIVE MANAGEMENT OF PRIMARY DISEASE

Lymph nodes are also analyzed intraoperatively for metastatic disease in patients with certain malignancies to determine whether operative management of the primary disease should be pursued. These scenarios involve a number of clinical situations in which metastatic disease is not considered treatable by surgery.

Intraoperative Evaluation of Lymph Nodes in Non-small Cell Lung Cancer

One such example is found in the staging and treatment of lung cancer. Operative management of non-small cell lung cancer (NSCLC) includes surgical removal of the primary tumor by wedge resection, lobectomy, or even pneumonectomy. However, all of these interventions have been regarded as contraindicated, if the mediastinal (N2) lymph nodes are involved by metastatic disease. As a result, mediastinoscopy is often performed prior to initiating procedures directed at the removal of the primary tumor. If frozen section analysis of the excised lymph nodes proves negative, the surgeon may proceed to the resection of the primary tumor, sparing the patient from the risk and cost of a second operation and anesthesia [30]. On the other hand, if mediastinal lymph nodes are positive, the procedure is aborted, and a central line for chemotherapy can be placed while the patient is still anesthetized [30]. As with SLN biopsy in breast cancer, there are significant consequences to both false-positive and false-negative results in intraoperative evaluation of lymph nodes in patients with NSCLC. A false-positive result may lead to the procedure being aborted unnecessarily, necessitating a second procedure (with its attendant risks) after permanent sections have been evaluated. False-negative results have the potential to be even more damaging, as patients unnecessarily undergo interventions associated with significant complications. In a systematic review of studies considering the accuracy of cervical mediastinoscopy, no false positives were reported in a total of 20 studies enrolling 6,505 patients [31]. Frozen sections yield few false negatives with sensitivities reported as 92.2% [32] and 94.5% [33] in two studies of open mediastinoscopy for bronchogenic carcinoma, which compares favorably with the 68–88% sensivity reported for non-invasive modalities [34–37]. Recent reviews suggest that unanticipated N2 disease can still be managed operatively if complete resection with negative margins can be achieved [38], which could limit use of frozen sections in this setting. For now, however, this application meets the requirement that a frozen section result may lead to real-time changes in management.

Interpretation is complicated by other potential etiologies of mediastinal lymphadenopathy in patients with lung adenocarcinoma. In addition to metastatic disease, the pathologist must consider a broad differential diagnosis, including lymphoma, sarcoidosis, and infectious etiologies, such as tuberculosis [30]. Appropriate precautions should be taken with these samples, and one should be prepared to submit tissue for bacteriologic studies or get a "lymphoma work-up" as discussed in previous chapters, as needed. However, when neoplastic epithelial cells are present, the diagnosis of metastatic disease is confirmed [39]. Finally, it is notable that despite the impressive accuracy of mediastinoscopy, endobronchial ultrasound (EBUS)-guided transbronchial fine needle aspiration is emerging as a less invasive approach [40].

Intraoperative Evaluation of Lymph Nodes in Pancreatic Cancer and Prostate Cancer

Lymph node frozen section evaluations are also potentially informative in cases of pancreatic cancer. As with NSCLC, operative management is indicated in the management of pancreatic adenocarcinoma, if disease is limited to an anatomic region that can be removed by en bloc resection. However, if the involvement of lymph nodes beyond the peripancreatic area is detected, a procedure with curative intent may be aborted in lieu of a palliative bypass operation [41]. In a study that combined all metastatic sites, including lymph nodes, the accuracy of intraoperative diagnosis was 97% [41]. Hence, pancreatic adenocarcinoma (like lung carcinoma) represents a case where intraoperative consultation can meet the fundamental criterion of altering operative management.

Historically, prostate adenocarcinoma has represented a third situation in which intraoperative assessment of lymph nodes was often requested. In patients with organ-confined prostate cancer, radical prostatectomy may be indicated. However, a number of significant risks exist with this procedure, including irreversible incontinence and sexual dysfunction. As a result, radical prostatectomy is contraindicated in patients with pelvic lymph node metastases, because there has been no demonstrable survival benefit to radical prostatectomy in these patients. Not surprisingly, frozen sections were frequently requested on pelvic lymph nodes in patients who are otherwise candidates for radical prostatectomy. A meta-analysis performed in the mid-1990s found a respectable composite sensitivity of 66.5% [42]. Given the low number of patients with lymph node involvement, frozen sections performed on all patients would have spared 7% of patients an unnecessary radical prostatectomy [42]. In other words, at that time, frozen sections had to be performed on pelvic lymph node dissections from approximately 14 to 15 patients to avoid a single unnecessary radical prostatectomy.

Since the mid-1990s, the natural history of prostate cancer has changed considerably. Screening with prostate serum antigen (PSA) and annual digital rectal examinations have become more commonplace, leading to an increase in the number of cases of organ-confined prostate cancer [43]. The higher proportion of patients with organ-confined disease has further decreased the clinical utility of intraoperative consultation, as lower percentages of patients with carcinoma had involved lymph nodes. Meanwhile, algorithms based on preoperative data that can predict risk of lymph node involvement have also been validated. Utilizing PSA level, Gleason score, and clinical stage, these prediction models categorize tumors as low (PSA less than 10, Gleason score less than 7, and clinical stage T1 or T2), intermediate (PSA 10–20 or Gleason score = 7 and T1 or T2 clinical stage), or high risk (PSA greater than 20, or Gleason score greater than 7 or clinical stage T3 or T4). Patients with low-and intermediate-risk tumors have a less than 1 and 4% risk of lymph node involvement, respectively [43]. As a result, the yield of lymph node frozen sectioning is too low to justify routine use of this procedure in low- and intermediate-risk patients. The value of intraoperative consultation in patients with high-risk disease is more controversial. Studies have not consistently agreed on the clinical value of lymph node frozen sections in these individuals, possibly due to methodological variability between studies [44].

MALIGNANT MIMICS OF CARCINOMA

Lymphoma

The possibility that a malignancy other than carcinoma is present in a lymph node may be overlooked, particularly for cases in which a diagnosis of carcinoma has been previously established. In cases without a prior diagnosis of malignancy, the surgical pathologist often thinks primarily of carcinoma when assessing lymph nodes for tumor. However, on occasion, it is not metastatic disease, but rather lymphoma that is the cause of the patient's lymphadenopathy. Obviously, the histologic findings vary depending on the specific type of lymphoma with some lymphomas being more prone to causing diagnostic confusion. For instance, anaplastic large-cell lymphoma may primarily involve the lymph node as large epithelioid cells with a somewhat cohesive nature filling the nodal sinuses, features that may resemble both melanoma and poorly differentiated carcinoma (Fig. 5.1). Poorly differentiated carcinoma with monotonous cytology, scant cytoplasm, and absence of clear-cut gland formation or squamous differentiation can also occasionally be confused with large B-cell lymphoma. Another scenario, in which lymphoma and carcinoma may be confused, is the rare instance of lymphoma showing "signet ring cell" differentiation (Fig. 5.2), which can be confused with metastatic carcinoma of various sites, particularly lobular carcinoma of breast or poorly differentiated primary gastrointestinal tumors. Thankfully, these potential pitfalls are not at all common, as most instances would require that a patient have an unrecognized hematolymphoid malignancy in addition to the solid tumor for which the lymph node is being assessed by frozen section.

In the rare circumstance that this situation presents itself, several diagnostic pearls may prove helpful during intraoperative consultation [45]. Although the exceptions noted above do exist, carcinoma

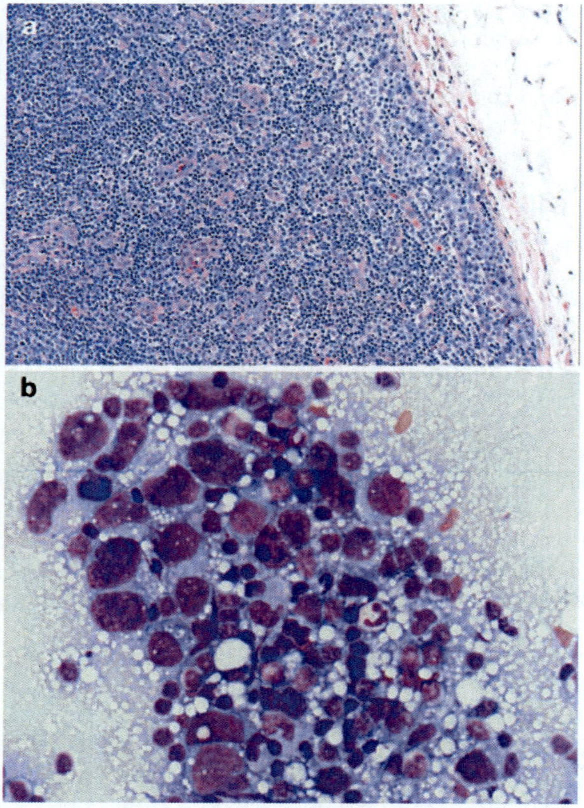

FIGURE 5.1 Anaplastic large-cell lymphoma with prominent sinusoidal involvement. Permanent section slides of a lymph node involved with anaplastic large-cell lymphoma, predominantly present within the subcapsular sinus area. The large cells often have abundant eosinophilic cytoplasm and can be confused with metastatic carcinoma. (**a**) Scanned H&E-stained slide, original magnification, ×20. Intraoperative touch preparation of lymph node involved by anaplastic large-cell lymphoma showing vague clustering of the large epithelioid cells, which can also lead to confusion with carcinoma. (**b**) Romanowski stain, ×60.

tends to produce a distinct pattern of nests or islands, with intervening strands of normal lymphatic tissue, which is distinct from what is observed in most lymphomas. Other features of the pattern of growth may be distinctive. Carcinoma usually respects the lymph node capsule. Tumor often fills the subcapsular sinus and does not

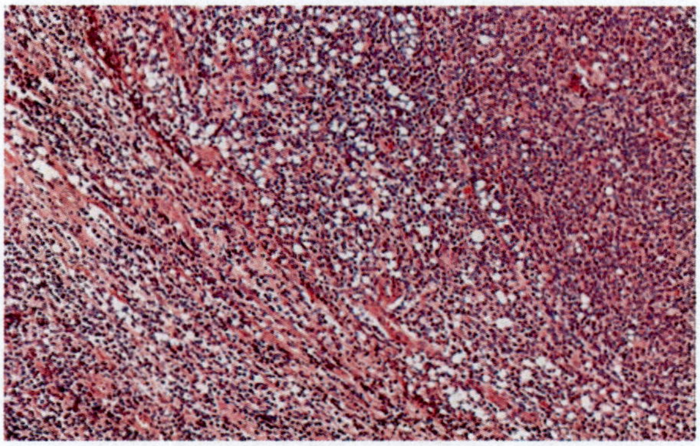

FIGURE 5.2 Follicular lymphoma with signet ring cells. Permanent section slide of a lymph node with grade 1 follicular lymphoma with signet ring cell morphology. The abnormal signet ring cells in this case are B cells that share the phenotype of the neoplastic follicles (scanned H&E-stained slide, original magnification, ×20).

transgress the capsule, while lymphoma frequently infiltrates surrounding soft tissue without regard for the lymph node capsule. If distinction of poorly differentiated carcinoma from lymphoma is a consideration, assessment of touch preparation cytology may prove helpful, in that the cohesive nature of carcinoma cell groups is generally readily distinguished from the noncohesive nature of lymphoma populations. Exceptions to this occur in both directions. For instance, the neoplastic cells of anaplastic large-cell lymphoma may form clusters, and in lobular or signet ring carcinomas, the neoplastic cells may not form cohesive groups.

Melanoma

Metastatic melanoma in lymph nodes may be confused with metastatic carcinoma. Similar to lymphoma, the likelihood of this occurring in the selected settings of lymph nodes submitted for frozen section analysis is low. The chances of this error are increased in cases of amelanotic melanoma or in the absence of a known primary lesion. Melanoma may mimic a wide variety of other neoplasms including sarcomas, lymphomas, or carcinomas, as the cytologic features vary considerably from case to case (Fig. 5.3). The presence of dusty brown pigment in the cells of interest may be a clue to the nature of the malignant cells. True glandular or squamous differentiation also allows distinction of carcinoma from melanomas in most cases. Similar to what is seen in lymphoma, touch preparations

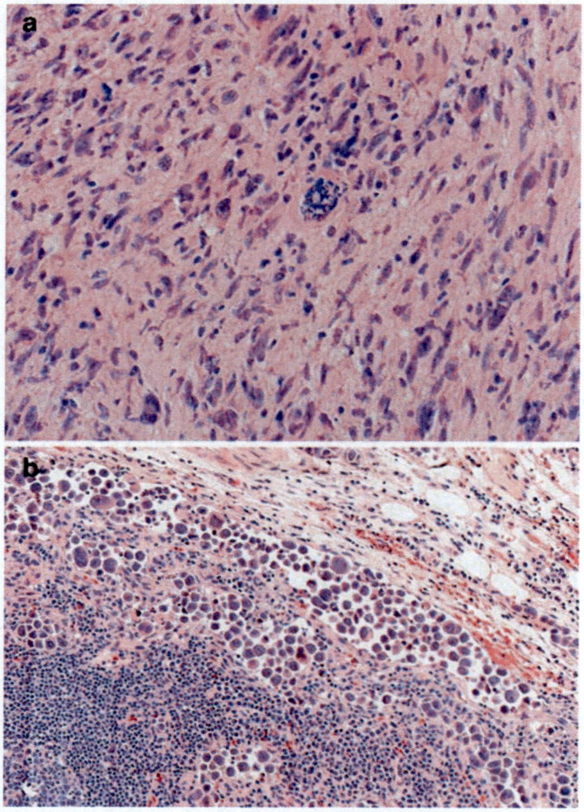

FIGURE 5.3 Different morphologies of metastatic melanoma. Actual frozen section slide showing a sarcomatoid variant of metastatic melanoma with focally spindle cell morphology with high-grade nuclear features and bizarre mitotic figures [(**a**) H&E, ×40]. A permanent section slide of metastatic melanoma shows a discohesive cell population filling and distending the sinuses, potentially mimicking anaplastic large-cell lymphoma and carcinoma [(**b**) scanned H&E-stained slide, original magnification ×20].

may be helpful in separating the cohesive cell groups typical of carcinoma from the noncohesive population of melanoma.

Soft-Tissue Tumors

Rare soft-tissue tumors (sarcomas) may be found in lymph nodes either primarily or as metastatic tumors. Primary lymph node sarcomas are exceedingly rare and because sarcomas of other sites frequently metastasize via hematogenous routes, their metastases are likewise uncommonly found in lymph nodes. These tumors can assume a wide variety of appearances, including some that mimic metastatic carcinoma, but often a prominent spindle cell component is present, with a fascicular or storiform arrangement of the cells rather than gland formation. Some sarcomas typically have an epithelioid appearance, and in certain instances, such as synovial sarcoma, there may be a true epithelial component present, which could lead to erroneous classification as carcinoma. Occasional cases of melanoma (Fig. 5.3a) or carcinoma may have a distinctly spindle cell or sarcomatoid appearance. Distinction of these lesions on frozen section may not be possible or even necessary. A diagnosis of "undifferentiated spindle cell neoplasm, pending further evaluation" rendered at the time of frozen section may be sufficient in such cases. If the differential diagnosis between carcinoma and sarcoma is considered at the time of frozen section analysis (often in cases with no prior diagnosis of malignancy or in pediatric patients), submission of fresh material for cytogenetic analysis may prove invaluable.

TECHNICAL ASPECTS

Intraoperative assessment of lymph nodes can be a difficult task. Section quality tends to be inferior to that available on permanent sections of formalin-fixed, paraffin-embedded tissue (Fig. 5.4). In addition, small foci of disease can easily be missed due to plane of section or misinterpretation of histologic findings.

Artifacts of Frozen Section

Frozen section interpretation may be compromised by the poor quality of the tissue sections themselves. Artifacts of preparation can make interpretation difficult. Specific problems include irregular staining, tearing, folding, venetian blind effects, etc., all of which occur more commonly with frozen sections than with permanent sections and can complicate the pathologist's task (see Fig. 5.4) [3]. Lymph nodes with extensive fatty replacement, in particular, may present challenges in terms of section preparation. In spite of these architectural limitations, if the frozen section slides are promptly fixed, cytologic detail can be relatively well preserved and may lead to the correct diagnosis. If the sections are not promptly fixed and stained, other artifacts introduced in the preparation and processing of frozen sections may limit the cytologic assessment as well. In particular, air-drying artifacts may distort nuclear:cytoplasmic ratios and produce a loss of

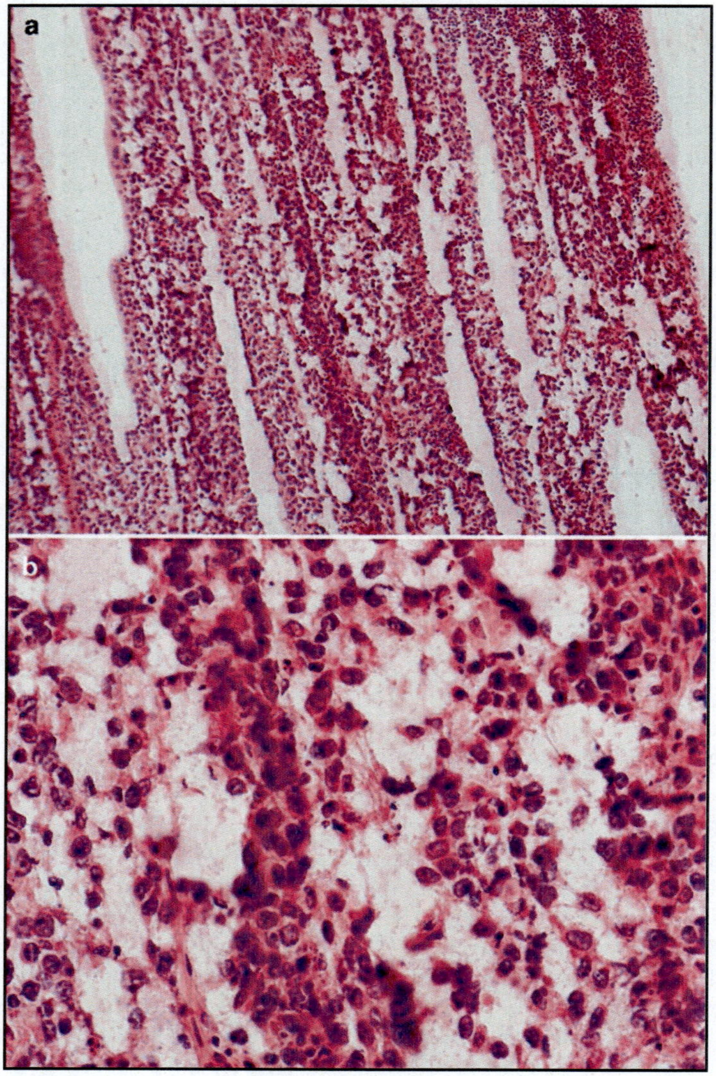

FIGURE 5.4 Artifacts of frozen sectioning. Hematoxylin- and eosin-stained actual frozen section slides and corresponding permanent section slides are shown here. Frozen sections of thymic carcinoma at low magnification (H&E, ×10) show extensive tearing and folds caused by knife chatter, resulting in areas of uneven staining (**a**); however, despite these limitations, cytologic detail can be appreciated at high magnification (H&E, ×40).

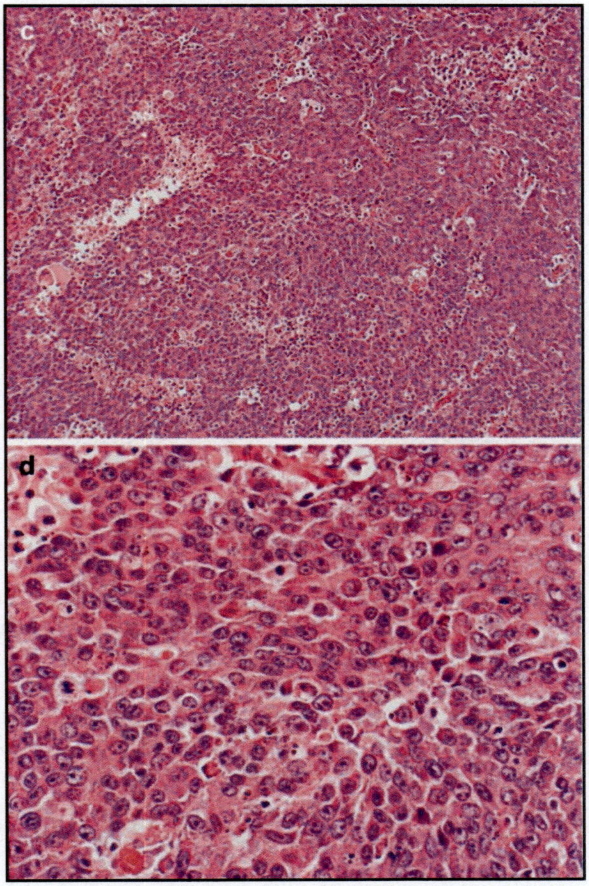

FIGURE 5.4 (continued) (**b**) Permanent section slides at low (H&E, ×10) and high (H&E, ×40) magnifications (**c**) and (**d**) lack the knife-chatter artifact, tissue folds, and tears and show well-preserved cytologic features.

chromatin detail. Because poor section quality may ultimately prove uninterpretable, it is important to have skilled staff and consistent methods of frozen section preparation to minimize artifacts. If poor sections are anticipated on the basis of fat content or extent of necrotic tissue, touch preparations may be an acceptable alternative. However, one may not consider this option necessary until the poor quality of sections is revealed, at which time it is too late to prepare touch preparations. Other artifacts, such as crush or thermal/cautery artifact, may lead to uninterpretable histology but are beyond the control of the pathologist. When reporting results, the limitations of such pre-laboratory artifacts should be mentioned.

Sampling Issues

The most straightforward cause of error at the time of frozen section is sampling error, which leads to false-negative interpretations, because tumor is present in the lymph node but not present in the histologic sections evaluated at the time of frozen section. Clinical suspicion may be increased by lymph nodes that are large, firm, lobulated, or otherwise abnormal, but these lymph nodes may be negative for metastatic disease. Conversely, even a grossly normal lymph node may be positive. Reporting negative results in such cases are technically "errors," although the inaccurate result is not an indictment of the diagnostician. Sampling error is almost an inevitable source of false-negative results as one evaluates sections of tissue that are several microns in thickness from specimens that may measure several centimeters in greatest dimension. Metastatic foci are often small and may not be grossly visible. The only detectable areas of metastasis may be restricted to deeper regions in the tissue sections that are not analyzed intraoperatively. Certain steps can be taken to reduce one's chances of missing a lymph node metastasis due to sampling error. In general, the entire lymph node should be sectioned perpendicular to the long axis of the lymph node into thin, 2–3 mm sections [9]. Depending on the size of the lymph node and gross appearance of the cut surface, the entire lymph node or representative sections of the lymph node may be frozen and sectioned. Careful gross examination of the cut sections of the lymph node should be performed, looking for areas of discoloration, hemorrhage, fibrosis, or necrosis, as these may be clues to the presence of metastatic disease. If the cut sections of the lymph node are grossly unremarkable, the entire lymph node should be frozen to fully evaluate the specimen for microscopic foci of metastatic disease. Although the specific number of slides prepared from each block of frozen tissue varies from institution to institution, in many places two levels are evaluated. Even so, small foci of metastatic disease will sometimes be missed, requiring revision of the final diagnosis once nodes are more fully analyzed on permanent sections. While a subset of these "false negatives" represents cases of micrometastatic disease, which may not always be prognostically equivalent to larger metastases, clinically relevant macrometastases may be missed as well.

Sectioning and Staining

Sections should be prepared on the microtome at 4–5 μm in thickness and mounted on glass slides. Once tissue is on the slide, immediate fixation in a formalin solution is necessary to prevent air-drying artifacts, which can make nuclei appear enlarged and obscure chromatin detail. The most frequent staining method is hematoxylin and eosin, which takes approximately 2 min to complete. It is noteworthy that depending on staining solutions used, some cells (such as eosinophils) do not have the same staining properties in frozen section as in permanent section slides. Cover slips are mounted on the slides with one or two drops of a suitable mounting medium with care to avoid entrapment of air bubbles, which may hinder microscopic evaluation.

Imprint Cytology

In cases where the lymph node appears grossly positive, imprint cytology with hematoxylin and eosin or a type of Romanowski stain (Fig. 5.5) may be attempted in place of frozen section analysis. This method is faster, induces fewer artifacts, and does not waste tissue.

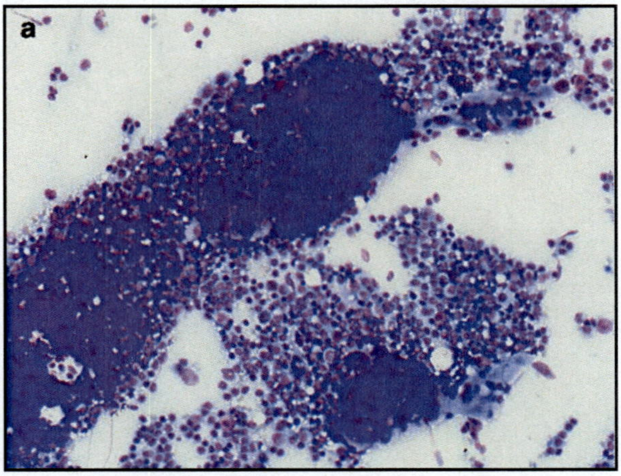

FIGURE 5.5 Lymph node imprint cytology. Intraoperative imprint cytology of lymph nodes can also be used to confirm the diagnosis of carcinoma. These examples highlight the clustering of carcinoma cells from a patient with endometrioid adenocarcinoma in a Romanowski-stained air-dried preparation [(**a**) ×10]. Higher magnification images of touch preparations highlight the differences between Romanowski-stained air-dried touch preparations [(**b**) ×60] and hematoxylin- and eosin-stained formalin-fixed touch preparations [(**c**) ×60].

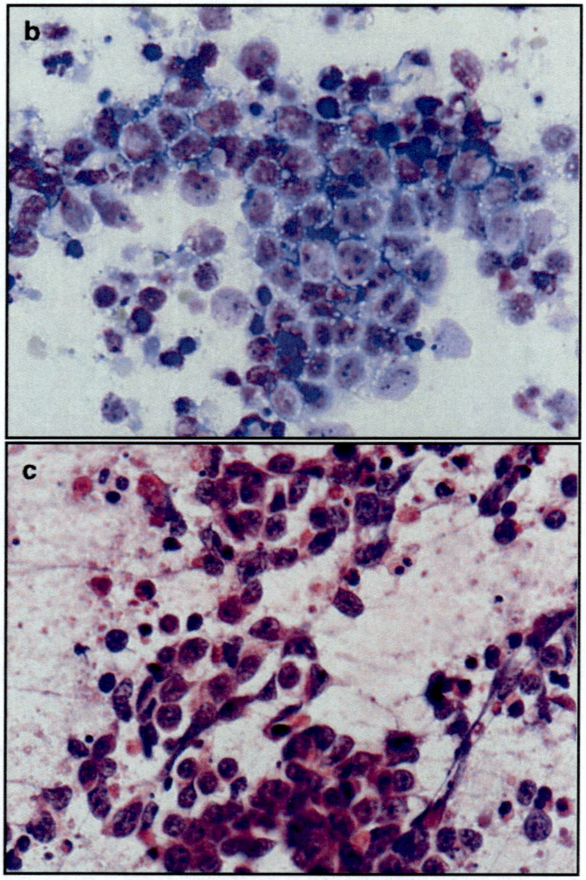

FIGURE 5.5 (continued)

Touch imprints are prepared simply by touching a glass slide to the cut surface of the tissue of interest. In some instances, the tissue may need to be blotted dry for best results (particularly for bloody tissue or small tissue fragments submitted in saline solution). One difficulty that may be encountered is that the tissue fragments adhere to the glass slide. This can sometimes be avoided if the tissue is placed on a paper towel before preparing the touch impression, since the tissue will adhere more tightly to the paper towel. (It may be necessary to moisten the paper towel before removing the tissue from it to avoid tearing the tissue.) If a Romanowski stain is to be used, the slide is air-dried and stained, but if an H&E stain is going to be prepared, the slide must be fixed immediately in formalin to avoid air-drying artifacts. It should be noted that since imprint cytology lacks corresponding architectural features, this method is suboptimal for determining the size of the metastatic focus and for delineation of extracapsular extension.

Assessing the Sections

Once sections are prepared, an organized and systematic approach to assessing tumor should be employed. Being familiar with the clinical history (i.e., the size or extent of the primary tumor and history of prior therapy) may be useful. Knowing the pathologic details of the primary tumor, particularly the histologic type and grade, also gives the frozen section pathologist insight into what is being sought in the lymph node. Probably, the most useful method of checking the "history" is comparison with prior diagnostic slides (such as from a needle biopsy of the breast or even cytology preparations). Because the large majority of lymph node metastases closely resemble their primary counterparts, having such sections available for comparison prior to or at the time of frozen section analysis can prove invaluable. It should be remembered, however, that metastases are not always identical to the primary tumor, so it is good to keep an open mind when evaluating the frozen section slides.

One must begin with a basic understanding of normal lymph node histology and cytologic features of cells in the various compartments of the lymph node. It is important to know that single large cells (such as transformed lymphocytes/immunoblasts) can be identified in the capsular sinuses or in the paracortex in reactive lymph nodes. These cells must not be interpreted as isolated metastatic carcinoma cells. Similarly, there may be nodules composed of medium to large cells exhibiting a high nuclear:cytoplasmic ratio and mitotic figures or apoptotic cells. Depending on the location in the lymph node and associated cell populations or architectural features, these groups of large cells can usually be recognized as reactive germinal centers (Fig. 5.6) or occasionally populations of plasmacytoid dendritic cells. Collections of histiocytes/macrophages in the lymph node sinuses or lymph node parenchyma may appear cohesive with an epithelioid appearance (Fig. 5.7a). Again, careful attention to cytologic detail (low nuclear:cytoplasmic ratio, small to medium sized nuclei with regular chromatin distribution and inconspicuous nucleoli, and in some locations, coarse pigmentation) will help distinguish these normal constituents of the lymph node from groups of malignant cells (Fig. 5.7b).

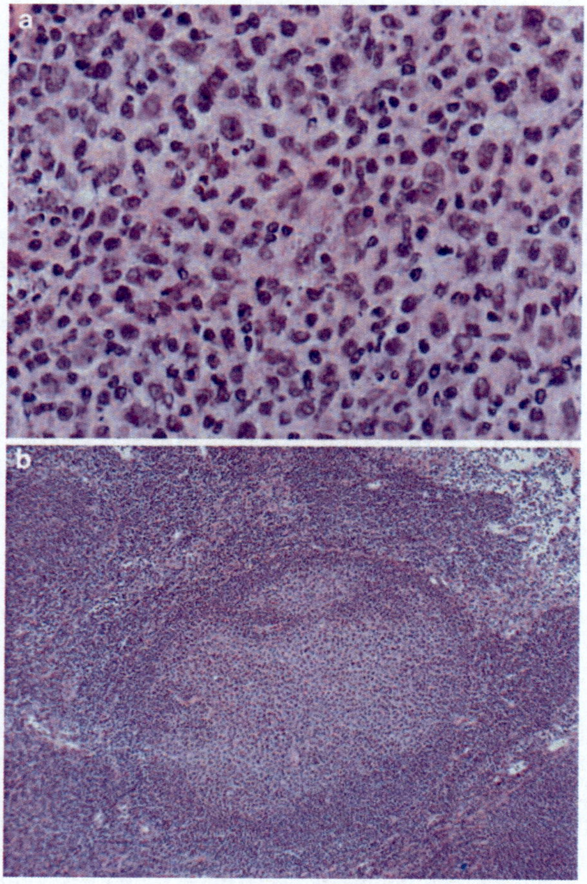

FIGURE 5.6 Reactive germinal center. This actual frozen section slide shows that reactive germinal centers may cause concern due the number of large lymphoid cells present at high magnification [(**a**) H&E, ×60]. However, assessment at lower magnification confirms the associated population of benign small mantle zone lymphocytes typical of reactive follicles [(**b**) H&E, ×10].

An approach to the evaluation of lymph nodes, partly based on the biology of metastasis, may also prove beneficial. During the earliest stages of lymph node involvement, metastatic tumor cells arrive via afferent lymphatics, which empty into subcapsular sinuses. These and other lymph node sinuses may become

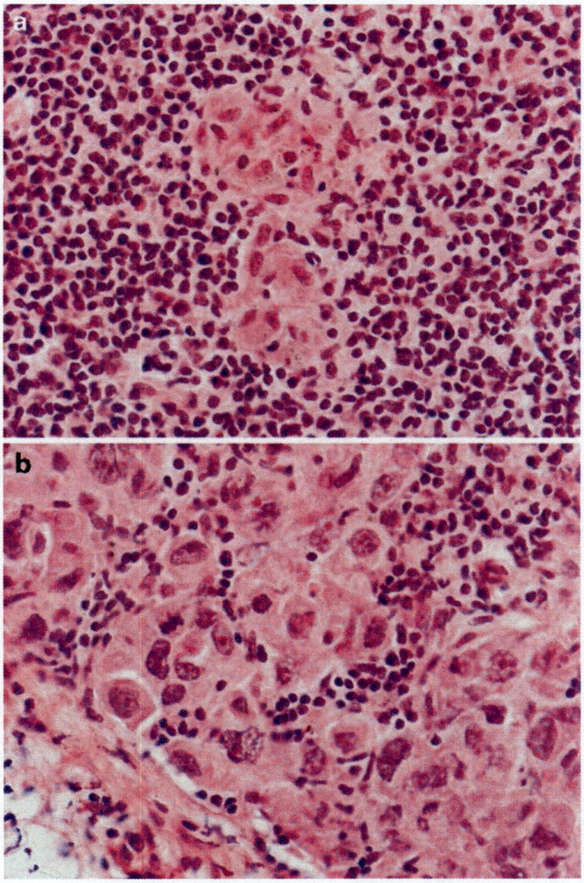

FIGURE 5.7 Cytologic distinction between sinus histiocytes and metastatic carcinoma. Actual frozen section of lymph nodes containing clusters of epithelioid macrophages (histiocytes) [(**a**) H&E, ×60] may cause some concern as these cells may be two to three times the size of a resting small lymphocyte. However, when compared with the metastatic lung cancer also present in the same lymph node, the marked degree of cytologic atypia, nuclear pleomorphism, and mitotic activity makes the distinction between these populations clear [(**b**) H&E, ×60].

distended by a cellular infiltrate (Fig. 5.8). Only later does the tumor spread to the medulla and cortex, progressively replacing the parenchyma and eventually effacing the normal nodal architecture. Given this natural history, the pathologist may pay

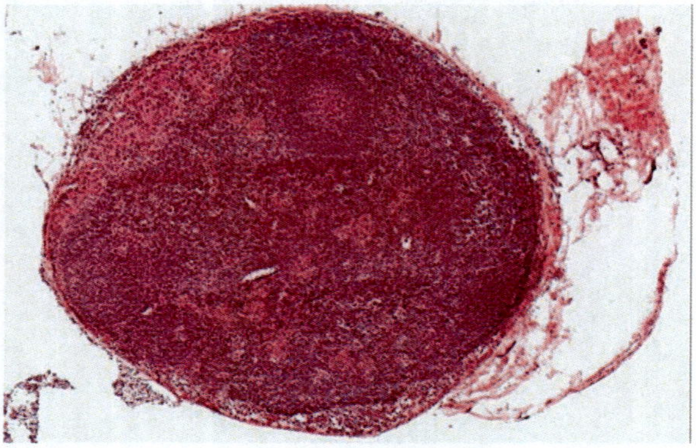

FIGURE 5.8 Prominent subcapsular sinus involvement by metastatic carcinoma. This actual frozen section slide shows metastatic squamous cell carcinoma involving and expanding the subcapsular sinus with early infiltration of the lymph node parenchyma (scanned slide; original magnification ×20).

special attention to the subcapsular sinuses, particularly in benign-appearing lymph nodes, where small, early areas of disease could be easily missed (Fig. 5.9) [45].

Although the subcapsular sinus is the most frequent site of early involvement of lymph nodes, in a given histologic section, small foci of carcinoma may be present only in nonsubcapsular areas (Fig. 5.10); so careful assessment of all tissue on the slide is essential in all cases. Other possible patterns of involvement include a general sinusoidal pattern mimicking sinus histiocytosis (Fig. 5.11), extensive or focal parenchymal involvement without involvement of sinuses, or even a nodular pattern that, at first blush, may mimic reactive lymphoid follicles or granulomata (Fig. 5.12). As mentioned previously, careful attention to cytologic and architectural features of the cell populations in question allows distinction of metastatic disease from normal or reactive lymph node features in the large majority of cases. For cases in which a diagnosis of carcinoma has not been previously established, it is important to remember that other tumor types, such as melanoma and anaplastic large-cell lymphoma as described above, may present with a prominent sinusoidal distribution.

Metastatic foci of carcinoma, in which the tumor cells produce well-formed glands or obvious keratinization, are readily recognized on frozen sections or on touch preparations. Similarly, metastatic foci comprising sheets of cells with high-nuclear grade

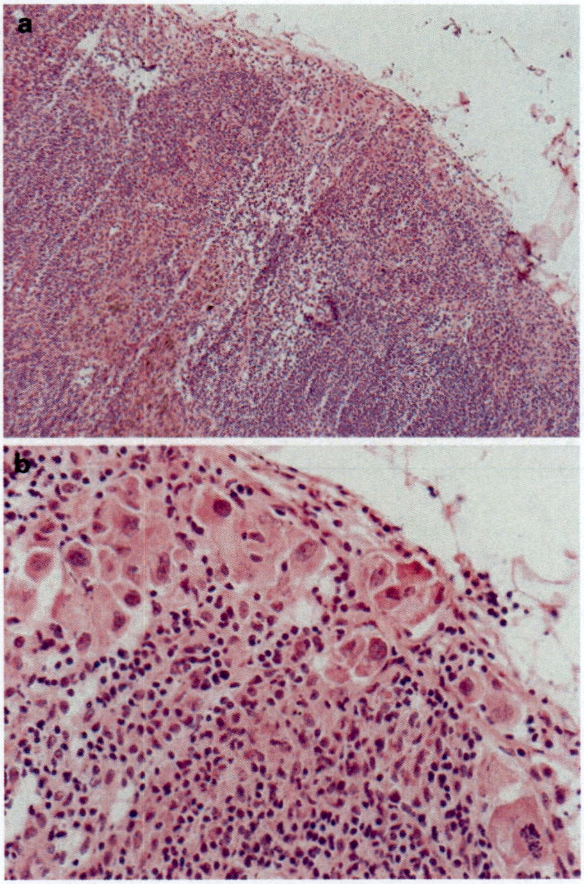

FIGURE 5.9 Subtle subcapsular sinus involvement by metastatic carcinoma. In this actual frozen section slide of a mediastinal lymph node, which also shows metastatic squamous cell carcinoma, the subtle involvement of the subcapsular sinus, could be overlooked at scanning magnification [(**a**) H&E, ×10]. However, higher magnification [(**b**) H&E, ×40] again makes the neoplastic nature of the cells in the subcapsular sinus quite obvious.

or with a brisk desmoplastic stromal response are generally easy to detect at frozen section. However, not all foci of metastatic disease are as easily detected on frozen section slides. Microscopic foci of tumor may be hidden by artifacts such as tissue folds or tears, and in some cases, the metastatic disease can produce very subtle patterns of involvement that can be difficult to recognize even on

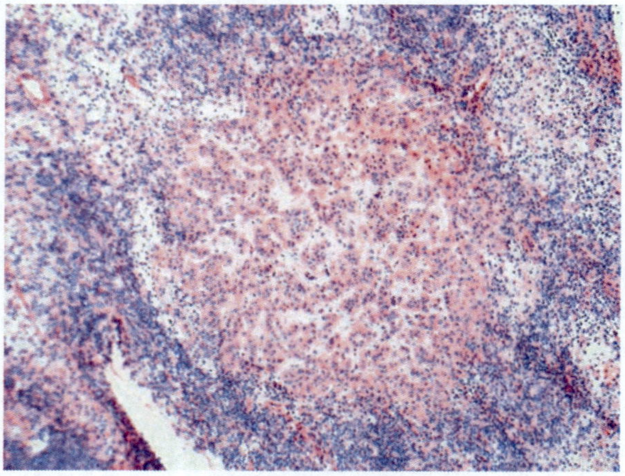

FIGURE 5.10 Metastatic carcinoma involving the lymph node parenchyma. In this actual frozen section slide, the micrometastatic focus of ductal carcinoma of the breast is present deep within lymph node parenchyma. There was no subcapsular tumor present in the lymph node, underscoring the importance of evaluating the entirety of the tissue on the frozen section slide. Note the somewhat smudgy appearance to the staining in this section, which is a byproduct of failure to quickly fix the slide in formalin prior to staining (air-drying artifact) (H&E, ×10).

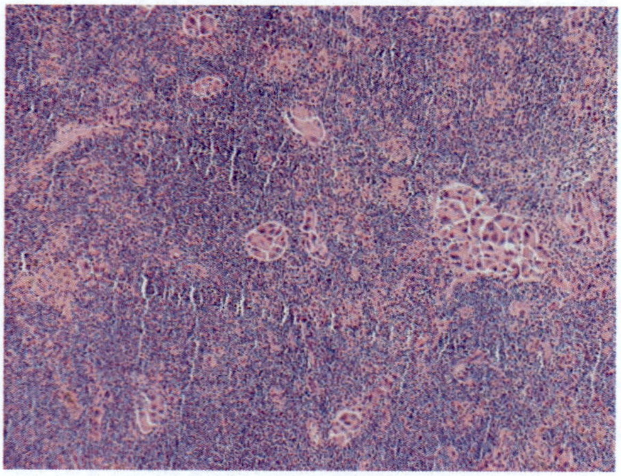

FIGURE 5.11 Metastatic carcinoma involving nonsubcapsular sinuses. This slide was prepared from the remnant of tissue that was subjected to frozen section analysis and shows involvement by metastatic carcinoma. The neoplastic cells are present in sinuses that are deep within the lymph node (nonsubcapsular sinuses). While this distribution of abnormal cells may overlap with that observed in sinus histiocytosis, the malignant cytologic features distinguish these cells from benign sinus histiocytes (H&E, ×10).

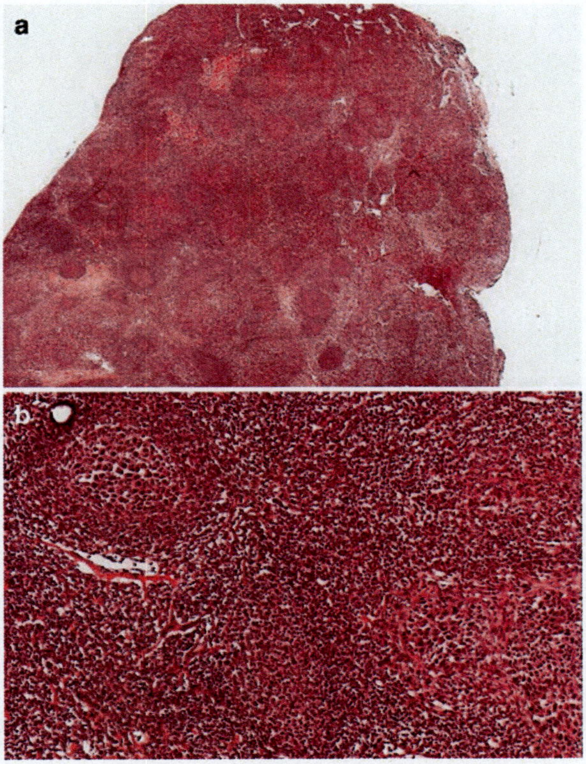

FIGURE 5.12 Metastatic carcinoma with a nodular growth pattern. In this lymph node, the scanning magnification images of the actual frozen section slide show a nodular pattern to the lymph node, reminiscent of follicular hyperplasia [(**a**) low-power image of scanned slide, original magnification ×20]. Many of the nodules have what appears to be a benign mantle zone. Higher magnification shows that the nodules are (in fact) a mixture of reactive lymphoid follicles and nodules of metastatic squamous cell carcinoma [(**b**) high-power image of scanned slide, original magnification ×20].

well-prepared sections. For instance, lobular carcinoma of breast or malignant melanoma may metastasize in a single cell pattern with relatively small cells. Depending on their level of differentiation and cytologic atypia, the metastatic tumor cells in such cases may closely resemble sinus histiocytes. Lobular carcinoma, in particular, may have a "shotgun" pattern of lymph node involvement producing a picture that resembles paracortical expansion with numerous "histiocytes" typical of that seen in dermatopathic

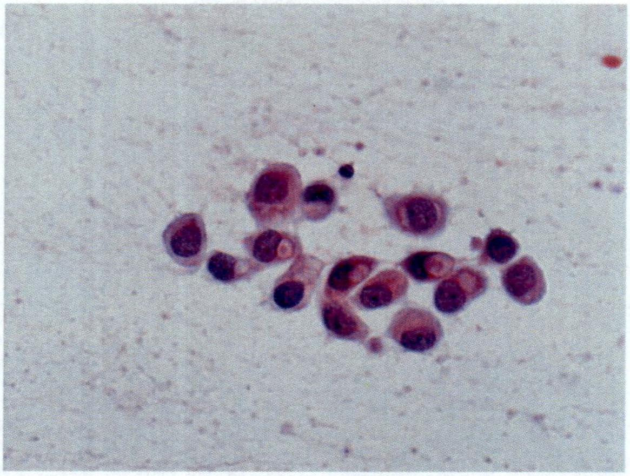

FIGURE 5.13 Imprint cytology of lymph node involved by lobular carcinoma. Formalin-fixed touch imprint cytology slide prepared from a sentinel lymph node in a patient with lobular carcinoma of the breast. The presence of intracytoplasmic lumina and prominent "targetoid" mucin help to distinguish these tumor cells from benign macrophages (H&E, ×60).

or viral lymphadenopathies. Although some cases will ultimately require immunohistochemical analysis to definitively identify the neoplastic population, attention to cytologic detail, such as noting the presence of intracytoplasmic vacuoles with characteristic targetoid mucin can help distinguish the lobular carcinoma cells from admixed macrophages. Imprint cytology may facilitate this assessment in cases where frozen section quality is limited (Fig. 5.13).

Activated high endothelial venules in the lymph node may have slight nuclear enlargement and nucleoli, mimicking individual well-formed glands of metastatic carcinoma with low-grade cytology, such as in grade 1 ductal carcinoma of the breast. These and other potential pitfalls confronted during frozen section analysis of lymph nodes are considered more fully in Chap. 6.

When lymph nodes are positive, additional characteristics of the diseased lymph node may be clinically significant. In breast carcinoma, the size of the metastatic focus detected may have an impact on management. For patients with micrometastatic disease (foci of carcinoma less than 2 mm in diameter) (Fig. 5.14), a complete axillary lymph node dissection may not be warranted, while the intraoperative detection of macrometastatic disease (Fig. 5.15) often

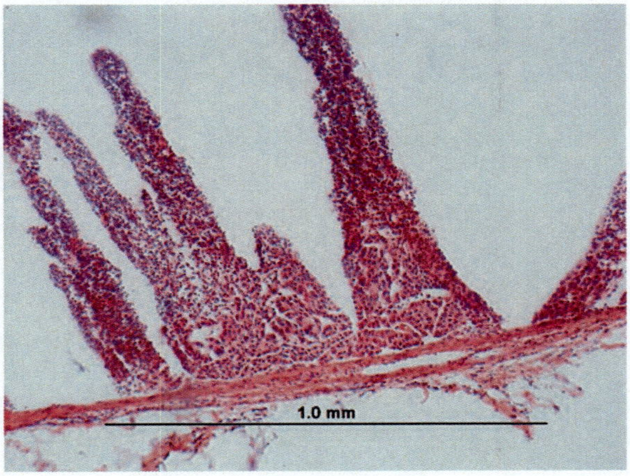

FIGURE 5.14 Micrometastatic focus of carcinoma in a sentinel lymph node. This minute focus of metastatic carcinoma, identified despite the extensive artifact evident in this actual frozen section slide is classified as a micrometastasis (measuring between 0.2 and 2 mm). This finding on a sentinel lymph node biopsy may no longer prompt a completion axillary lymph node dissection, a situation which may limit the utility of frozen section evaluation in this setting (H&E, ×20).

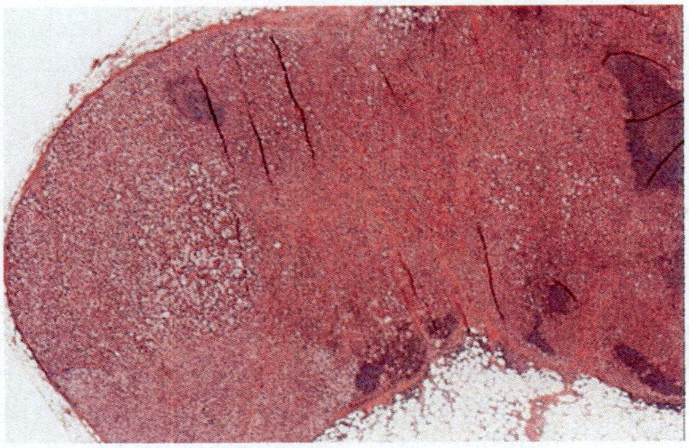

FIGURE 5.15 Macrometastatic focus of carcinoma in a lymph node. This scanned image of an actual frozen section slide shows a large expanse of tumor present in a sentinel lymph node. The presence of macrometastatic disease in axillary sentinel lymph nodes (greater than 2 mm in greatest dimension) will likely prompt a completion axillary dissection at the time of frozen section (low-power image of scanned slide, original magnification ×20).

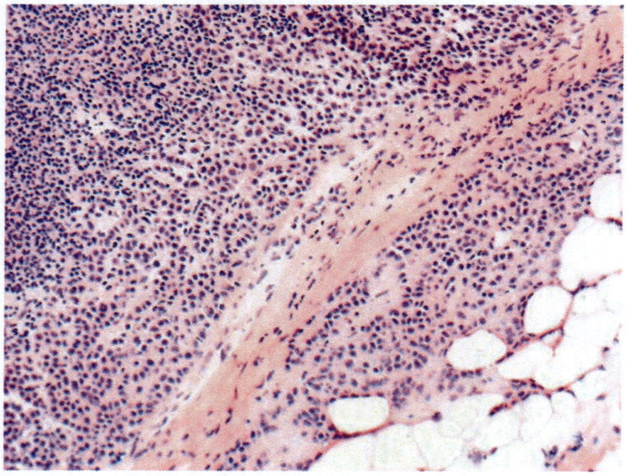

FIGURE 5.16 Extracapsular extension of metastatic carcinoma. Actual frozen section of a sentinel lymph node demonstrating extracapsular extension of metastatic infiltrating ductal carcinoma with invasion of perinodal fat (H&E, ×20).

leads to completion axillary lymph node dissection. As mentioned before, carcinoma that has spread beyond the capsule and into adjacent perinodal soft tissue may be an indicator that the tumor is not resectable and may lead to the surgeon aborting a planned primary tumor resection. It is important to recognize extranodal extension (Fig. 5.16) if it is present in frozen sections (even if it will not alter immediate surgical management), since it is considered a poor prognostic factor in many cancers. For instance, in women with breast cancer and positive SLNs, extranodal extension is an independent risk factor for the presence of additional positive non-SLNs [46–49]. Likewise, extranodal extension is associated with decreased survival in patients with NSCLC [50].

Finally, interpretation of lymph node frozen sections may be complicated in cases where patients are found to have a second malignancy in addition to the patient's known carcinoma. The potential combinations of neoplasms are innumerable, but the most frequent offenders are likely low-grade lymphomas such as follicular lymphoma or small lymphocytic lymphoma. The presence of one malignancy is not a reason to disregard the possibility of a second lesion in the lymph node. The routine approach to looking for metastatic disease may draw attention away from the lymphoid populations in the lymph node, particularly if metastatic disease is

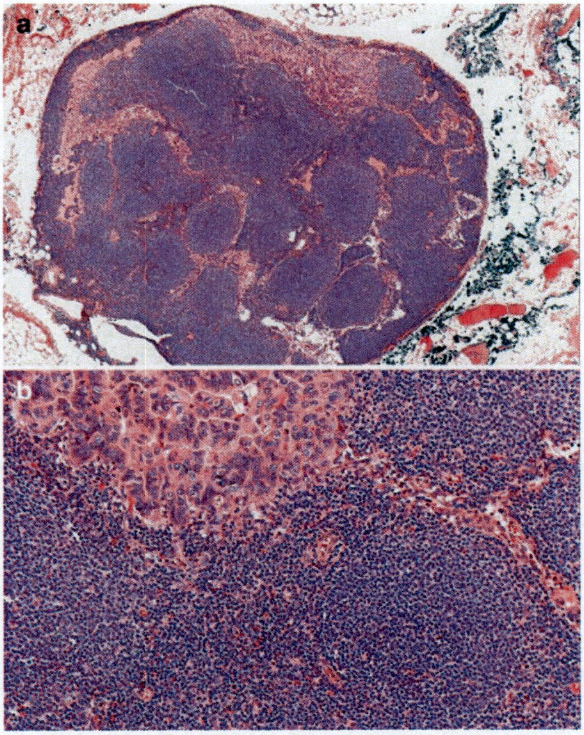

FIGURE 5.17 Metastatic carcinoma in a lymph node involved by follicular lymphoma, grade 1. Permanent section slide from a patient with gastric carcinoma showing metastatic disease. Of note in this case is the additional presence of abnormal closely packed follicles comprising almost exclusively centrocytes. Immunophenotyping by immunohistochemistry confirmed the neoplastic nature of these follicles [(**a** and **b**) H&E, ×4 and ×40].

identified (Fig. 5.17). However, the gross finding of a large, fleshy lymph node, which is particularly worrisome in older patients, who frequently have more fat replacement in their lymph nodes, may be the first indicator that something other than metastatic carcinoma may be present. On histologic sections, noting that the normal lymph node architecture is effaced with the absence of normal lymphoid follicles and distortion or obliteration of normal lymph node sinuses, noting closely packed follicles without well-formed mantle zones, or seeing a diffuse population of monotonous lymphoid cells may all be clues to the diagnosis of a lymphoproliferative disorder instead of or in addition to the presence of metastatic carcinoma.

WHAT SHOULD BE REPORTED?

In most instances, lymph nodes submitted for frozen section analysis are evaluated for the presence of metastatic disease, although other diagnostic possibilities are sometimes considered by the surgeon. Lymph nodes submitted for a clinical question of lymphoma are handled as previously discussed in Chaps. 2 and 3. An attempt should be made to answer the clinical question being asked. In many instances, a diagnosis of "Positive for metastatic carcinoma" or "No carcinoma identified" is sufficient to answer the question being posed. In some cases, during the discussion with the surgeon, other questions may arise (such as subtype of tumor, size of metastatic focus, etc.). Answering these questions as thoroughly as possible is advised, but the pathologist should note inherent limitations in the frozen section setting. For instance, it may not be possible to definitively determine the site of origin of a tumor based solely on morphology, although prior imaging studies or location of the lymph node may provide additional clues [45], and the size of a tumor focus measured on a glass slide should be regarded as a minimum estimate.

When other disease entities are recognized or suspected (i.e., granulomatous disease or some other cause of lymphadenopathy), these findings should also be reported to the surgeon, as it may lead to altered management. One example might be the procurement of additional tissue for microbiology cultures in cases where frozen section analysis for a suspected tumor reveals necrotizing granulomatous inflammation. Even noting the presence of marked follicular lymphoid hyperplasia may be correlated with a suspicious lymph node finding on positron emission tomography (PET) studies. In the end, it may be useful for the pathologist to specifically ask the surgeon at the time results are reported, if the reported findings are compatible with the clinical findings. If there is a significant discordance, additional sampling (and possibly additional frozen sections) may be warranted.

SUMMARY

The examples provided above illustrate the utility and some of the limitations of lymph node frozen sections in assessing metastatic malignancy. When applied to appropriate clinical questions, these intraoperative consultations can provide critical information on the status of lymph nodes draining a primary tumor or, in other cases, reveal the primary diagnosis. However, there are also a number of limitations to lymph node frozen sections. In addition to the clinical considerations and technical limitations discussed above, there are a number of potential pitfalls that may be encountered by the practicing pathologist asked to interpret sections from a frozen node. Some of these challenges are considered more fully in the following chapter.

REFERENCES

1. Ackerman LV, Ramirez GA. The indications for and limitations of frozen section diagnosis; a review of 1269 consecutive frozen section diagnoses. Br J Surg. 1959;46(198):336–50.
2. Laucirica R. Intraoperative assessment of the breast: guidelines and potential pitfalls. Arch Pathol Lab Med. 2005;129(12):1565–74.
3. Prieto VG. Use of frozen sections in the examination of sentinel lymph nodes in patients with melanoma. Semin Diagn Pathol. 2008;25(2): 112–5.
4. Seaman WB, Powers WE. Studies on the distribution of radioactive colloidal gold in regional lymph nodes containing cancer. Cancer. 1955;8(5):1044–6.
5. Gould EA, Winship T, Philbin PH, Kerr HH. Observations on a "sentinel node" in cancer of the parotid. Cancer. 1960;13:77–8.
6. Morton DL, Wen DR, Wong JH, et al. Technical details of intraoperative lymphatic mapping for early stage melanoma. Arch Surg. 1992;127(4):392–9.
7. Giuliano AE, Dale PS, Turner RR, Morton DL, Evans SW, Krasne DL. Improved axillary staging of breast cancer with sentinel lymphadenectomy. Ann Surg. 1995;222(3):394–9. discussion 399–401.
8. Giuliano AE, Kirgan DM, Guenther JM, Morton DL. Lymphatic mapping and sentinel lymphadenectomy for breast cancer. Ann Surg. 1994;220(3):391–8. discussion 398–401.
9. Schwartz GF. Clinical practice guidelines for the use of axillary sentinel lymph node biopsy in carcinoma of the breast: Current update. Breast J. 2004;10(2):85–8.
10. Flett MM, Going JJ, Stanton PD, Cooke TG. Sentinel node localization in patients with breast cancer. Br J Surg. 1998;85(7):991–3.
11. Van Diest PJ, Torrenga H, Borgstein PJ, et al. Reliability of intraoperative frozen section and imprint cytological investigation of sentinel lymph nodes in breast cancer. Histopathology. 1999;35(1):14–8.
12. Dixon JM, Mamman U, Thomas J. Accuracy of intraoperative frozen-section analysis of axillary nodes. Edinburgh breast unit team. Br J Surg. 1999;86(3):392–5.
13. Motomura K, Inaji H, Komoike Y, et al. Intraoperative sentinel lymph node examination by imprint cytology and frozen sectioning during breast surgery. Br J Surg. 2000;87(5):597–601.
14. Schwartz GF, Krill LS, Palazzo JP, Dasgupta A. Value of intraoperative examination of axillary sentinel nodes in carcinoma of the breast. J Am Coll Surg. 2008;207(5):758–62.

15. van de Vrande S, Meijer J, Rijnders A, Klinkenbijl JH. The value of intraoperative frozen section examination of sentinel lymph nodes in breast cancer. Eur J Surg Oncol. 2009;35(3):276–80.
16. Veronesi U, Paganelli G, Galimberti V, et al. Sentinel-node biopsy to avoid axillary dissection in breast cancer with clinically negative lymph-nodes. Lancet. 1997;349(9069):1864–7.
17. Dabbs DJ, Fung M, Landsittel D, McManus K, Johnson R. Sentinel lymph node micrometastasis as a predictor of axillary tumor burden. Breast J. 2004;10(2):101–5.
18. Colleoni M, Rotmensz N, Peruzzotti G, et al. Size of breast cancer metastases in axillary lymph nodes: clinical relevance of minimal lymph node involvement. J Clin Oncol. 2005;23(7):1379–89.
19. Cox CE, Kiluk JV, Riker AI, et al. Significance of sentinel lymph node micrometastases in human breast cancer. J Am Coll Surg. 2008; 206(2):261–8.
20. Gobardhan PD, Elias SG, Madsen EV, et al. Prognostic value of micrometastases in sentinel lymph nodes of patients with breast carcinoma: a cohort study. Ann Oncol. 2009;20(1):41–8.
21. Vanderveen KA, Ramsamooj R, Bold RJ. A prospective, blinded trial of touch prep analysis versus frozen section for intraoperative evaluation of sentinel lymph nodes in breast cancer. Ann Surg Oncol. 2008;15(7):2006–11.
22. Tew K, Irwig L, Matthews A, Crowe P, Macaskill P. Meta-analysis of sentinel node imprint cytology in breast cancer. Br J Surg. 2005;92(9): 1068–80.
23. Perez N, Vidal-Sicart S, Zanon G, et al. A practical approach to intraoperative evaluation of sentinel lymph node biopsy in breast carcinoma and review of the current methods. Ann Surg Oncol. 2005;12(4): 313–21.
24. Krishnamurthy S, Meric-Bernstam F, Lucci A, et al. A prospective study comparing touch imprint cytology, frozen section analysis, and rapid cytokeratin immunostain for intraoperative evaluation of axillary sentinel lymph nodes in breast cancer. Cancer. 2009;115(7):1555–62.
25. Fisher B, Jeong JH, Anderson S, Bryant J, Fisher ER, Wolmark N. Twenty-five-year follow-up of a randomized trial comparing radical mastectomy, total mastectomy, and total mastectomy followed by irradiation. N Engl J Med. 2002;347(8):567–75.
26. Giuliano AE, Hunt KK, Ballman KV, et al. Axillary dissection vs no axillary dissection in women with invasive breast cancer and sentinel node metastasis: a randomized clinical trial. JAMA. 2011;305(6):569–75.
27. Scolyer RA, Murali R, McCarthy SW, Thompson JF. Pathologic examination of sentinel lymph nodes from melanoma patients. Semin Diagn Pathol. 2008;25(2):100–11.
28. Ichikura T, Chochi K, Sugasawa H, et al. Individualized surgery for early gastric cancer guided by sentinel node biopsy. Surgery. 2006;139(4): 501–7.
29. Hartgrink HH, van de Velde CJ, Putter H, et al. Extended lymph node dissection for gastric cancer: who may benefit? Final results

of the randomized Dutch gastric cancer group trial. J Clin Oncol. 2004;22(11):2069–77.
30. de Montpreville VT, Dulmet EM, Nashashibi N. Frozen section diagnosis and surgical biopsy of lymph nodes, tumors and pseudotumors of the mediastinum. Eur J Cardiothorac Surg. 1998;13(2):190–5.
31. Detterbeck FC, Jantz MA, Wallace M, Vansteenkiste J, Silvestri GA, American College of Chest Physicians. Invasive mediastinal staging of lung cancer: ACCP evidence-based clinical practice guidelines (2nd edition). Chest. 2007;132(3 Suppl):202S–20.
32. Gephardt GN, Rice TW. Utility of frozen-section evaluation of lymph nodes in the staging of bronchogenic carcinoma at mediastinoscopy and thoracotomy. J Thorac Cardiovasc Surg. 1990;100(6):853–9.
33. Sanli M, Isik AF, Tuncozgur B, Akar E, Deniz H, Bakir K, Elbeyli L. The reliability of mediastinoscopic frozen sections in deciding on oncologial surgery in bronchogenic carcinoma. Adv Ther. 2008;25(5):488–95.
34. De Leyn P, Schoonooghe P, Deneffe G, et al. Surgery for non-small cell lung cancer with unsuspected metastasis to ipsilateral mediastinal or subcarinal nodes (N2 disease). Eur J Cardiothorac Surg. 1996;10(8): 649–54. discussion 654–5.
35. Deneffe G, Lacquet LM, Gyselen A. Cervical mediastinoscopy and anterior mediastinotomy in patients with lung cancer and radiologically normal mediastinum. Eur J Respir Dis. 1983;64(8):613–9.
36. Gdeedo A, Van Schil P, Corthouts B, Van Mieghem F, Van Meerbeeck J, Van Marck E. Comparison of imaging TNM [(i)TNM] and pathological TNM [pTNM] in staging of bronchogenic carcinoma. Eur J Cardiothorac Surg. 1997;12(2):224–7.
37. Ratto GB, Frola C, Cantoni S, Motta G. Improving clinical efficacy of computed tomographic scan in the preoperative assessment of patients with non-small cell lung cancer. J Thorac Cardiovasc Surg. 1990;99(3):416–25.
38. Detterbeck F. What to do with "surprise" N2? Intraoperative management of patients with non-small cell lung cancer. J Thorac Oncol. 2008;3(3):289–302.
39. Gordon IO, Suzue K, Husain AN. Lung, mediastinum, and pleura. In: Taxy JB, Husain AN, Montag AG, editors. Biopsy interpretation: the frozen section. 1st ed. Philadelphia, PA: Wolters Kluwer; 2010. p. 47–79.
40. Block MI. Transition from mediastinoscopy to endoscopic ultrasound for lung cancer staging. Ann Thorac Surg. 2010;89(3):885–90.
41. Cioc AM, Ellison EC, Proca DM, Lucas JG, Frankel WL. Frozen section diagnosis of pancreatic lesions. Arch Pathol Lab Med. 2002;126(10): 1169–73.
42. Davis GL. Sensitivity of frozen section examination of pelvic lymph nodes for metastatic prostate carcinoma. Cancer. 1995;76(4):661–8.
43. Beissner RS, Stricker JB, Speights VO, Coffield KS, Spiekerman AM, Riggs M. Frozen section diagnosis of metastatic prostate adenocarcinoma in pelvic lymphadenectomy compared with nomogram prediction of metastasis. Urology. 2002;59(5):721–5.

44. Taxy JB. Urinary tract (kidney, urothelium, and prostate). In: Taxy JB, Husain AN, Montag AG, editors. Biopsy interpretation: the frozen section. 1st ed. Philadelphia, PA: Wolters Kluwer; 2010. p. 118–48.
45. Ioachim HL. Tumor metastases in lymph nodes. In: Ioachim HL, Medeiros LJ, editors. Ioachim's lymph node pathology. Philadelphia, PA: Lippincott, Williams, and Wilkins; 2009. p. 590–8.
46. Changsri C, Prakash S, Sandweiss L, Bose S. Prediction of additional axillary metastasis of breast cancer following sentinel lymph node surgery. Breast J. 2004;10(5):392–7.
47. Fleming FJ, Kavanagh D, Crotty TB, et al. Factors affecting metastases to non-sentinel lymph nodes in breast cancer. J Clin Pathol. 2004;57(1):73–6.
48. Joseph KA, El-Tamer M, Komenaka I, Troxel A, Ditkoff BA, Schnabel F. Predictors of nonsentinel node metastasis in patients with breast cancer after sentinel node metastasis. Arch Surg. 2004;139(6):648–51.
49. Turner RR, Chu KU, Qi K, et al. Pathologic features associated with nonsentinel lymph node metastases in patients with metastatic breast carcinoma in a sentinel lymph node. Cancer. 2000;89(3):574–81.
50. Lee YC, Wu CT, Kuo SW, Tseng YT, Chang YL. Significance of extranodal extension of regional lymph nodes in surgically resected non-small cell lung cancer. Chest. 2007;131(4):993–9.

Chapter 6

Pitfalls of Frozen Section to Intraoperative Consultations of Evaluating Lymph Nodes for Involvement by Metastatic Malignancies: Benign Processes Mimicking Metastatic Carcinoma

INTRODUCTION

As discussed in Chap. 5, intraoperative frozen section evaluation of metastatic malignancy in lymph nodes is a technique that has the potential to guide clinical decisions in real time during a surgical procedure. One of the main advantages to this technique is that it allows clinicians to distinguish patients who may benefit from operative management from those who are unlikely to benefit. In general, the major applications of lymph node frozen section analysis include assessment of sentinel lymph nodes (SLNs) for metastatic disease to determine which patients may benefit from a more extensive lymph node dissection and the assessment of lymph nodes for metastatic disease to determine the appropriate management of a primary tumor. These major applications of lymph node frozen section are applied in distinct clinical settings. Although SLN assessment was initially developed for management of melanoma,

C.H. Dunphy (ed.), *Frozen Section Library: Lymph Nodes*,
Frozen Section Library, 10, DOI 10.1007/978-1-4614-1253-3_6,

its current use is primarily in patients with breast cancer. The intraoperative staging of lymph nodes may be used to determine if tumors are inoperable, because carcinoma has spread beyond the primary organ to lymph nodes. This use of frozen sections has been applied particularly to some lung, prostate, and pancreatic cancers. A more in-depth discussion of the indications for performing lymph node frozen sections was considered more fully in Chap. 5.

Although it is essential to render correct diagnoses at the time of frozen section, false-positive and false-negative results occur. Depending on the circumstances, a false-positive result may be more damaging than a false-negative result, or vice versa. For instance, a false-positive SLN, leading to completion axillary dissection, exposes patients to unnecessary risks of lymphedema, pain syndromes, or paresthesias, complications seen in up to 26% of women in one study [1]. In contrast, a false-negative result from a SLN biopsy might lead to an additional operative procedure for completion of the axillary dissection. In fact, only a subset of false-negative diagnoses would result in an additional unnecessary procedure, because in some instances an additional procedure would already be needed for re-excision of positive margins. In other situations, a false-negative result might lead to an unnecessarily morbid procedure being performed, such as a false-negative mediastinal lymph node that permits an extensive lung resection to be performed in a patient with metastatic disease. A false-positive frozen section result in this case might lead the surgeon to abandon the planned operation at that time and necessitate an additional surgical procedure at a later date to resect the primary tumor.

Fortunately, the reported incidence of false-positive results is quite low in most cases. Indeed, several series of intraoperative evaluation of SLN biopsies in patients with breast cancer have reported no false-positive results [2–7]. Nevertheless, there is no shortage of reports in the literature of falsely positive frozen sections, and it is likely that published series reflect some degree of publication bias, as institutions with a higher incidence of false-positive results are presumably less likely to report those findings.

The diagnosis of lymph node metastasis involves recognizing the presence of exogenous cells not normally found in the lymph node. Although this seems like a relatively straightforward task, there are several potential pitfalls of which the pathologist must be aware. Probably the biggest category of pitfalls is the false-positive diagnosis due to benign mimics of carcinoma (including epithelial and nonepithelial mimics). Lymph nodes do not usually harbor epithelial cell populations, but there are examples of epithelial or "epithelioid" cell populations that can be seen at frozen section and lead to a misdiagnosis of metastatic disease. These pitfalls may be due to benign epithelial inclusions (such as heterotopic glands) or nonepithelial mimics of carcinoma. In the former cases, normal epithelial structures may be found in the lymph node capsule, sinuses, or occasionally deeper in the parenchyma, where they may be mistaken for carcinoma. On the other hand, the nonepithelial mimics of carcinoma are a heterogeneous group of conditions, in which the implicated cells are not derived from an epithelium but take on an epithelioid appearance that can be misinterpreted on frozen section slides. Many of the individual examples in each category are themselves only rarely encountered. However, collectively, these mimics are not uncommon. Other potential frozen section pitfalls include true positive results, wherein an abnormal cell population is recognized but misclassified, and false-negative results (often due to sampling or failure to recognize small foci of neoplastic cells). Another form of pitfall may be related to incomplete or inaccurate clinical information for a given tissue fragment submitted for evaluation. The potential pitfalls and histologic characteristics that distinguish benign mimics of carcinoma from true metastases as well as other potential frozen section pitfalls are explored in the sections that follow.

EPITHELIAL INCLUSIONS

Epithelial inclusions represent cases in which heterotopic epithelia and glands are present in a lymph node. Because these inclusions represent true epithelial structures, which are not typically present in lymph nodes, they can easily be confused with metastatic adenocarcinoma. Epithelial inclusions can be derived from a wide variety of normal epithelia. In most cases, the ectopic cells are histologically related to the tissue normally drained by the lymph node. Here, these cases are referred to as "local heterotopic epithelial inclusions." A special situation is observed with Müllerian inclusions, in which uterine-type epithelium can be found in lymph nodes throughout the body. These elements are considered separately in the section entitled "Distant heterotopic epithelial inclusions." As discussed below, the distinction between benign inclusions and metastatic disease at the time of frozen section is in most cases based on the cytologic features. In some instances, the location of the inclusion may also be a useful feature in discriminating these two possibilities.

Local Heterotopic Epithelial Inclusions

Local heterotopic epithelial inclusions are benign epithelial elements rarely found in lymph nodes. In these cases, the ectopic tissue corresponds to epithelial structures from tissues normally drained by the lymph node in question. In many cases, this means that the inclusions represent benign tissue from the organ involved by the primary tumor, which makes these structure challenging mimics of carcinoma.

Epithelial inclusions have been seen in association with a number of organs, including salivary and thyroid glands, the pancreas, and squamous epithelium and mesothelium [8–10]. However, the greatest volume of literature has been published for cases of epithelial inclusions in the axillary nodes of women with breast cancer. Few comprehensive studies have been performed to evaluate the incidence of these inclusions, but in one series, 7 of 3,500 axillary SLNs contained heterotopic epithelium [11].

The histology of epithelial inclusions in axillary nodes overlaps with that observed in benign breast tissue. Three basic architectural forms have been identified. The first two consist of ducts and tubules or lobules, respectively. These elements are histologically similar to structures found in normal breast; the ectopic epithelium is a bilayer, composed of a luminal cuboidal-to-columnar epithelium and underling myoepithelial cells with associated basement membrane, both of which can be identified by immunohistochemistry [11, 12]. The third architectural form is a cystic variety, in which the lumen is lined by stratified squamous epithelium [11, 13–15].

Regardless of which architectural form is present, several histological features may allow the pathologist to reliably discriminate heterotopic epithelial inclusions in axillary nodes from metastatic adenocarcinoma [11, 12]. Unlike their malignant counterparts, heterotopic epithelial inclusions lack cytological atypia, have bland nuclear features, and do not elicit a desmoplastic response. Location can also be informative. Whereas metastatic disease is most often detected first in the subcapsular sinuses, benign inclusions are typically situated within the capsule. Evaluation can become more complicated when hyperplastic changes are present. Cases have been reported in which inclusions have features resembling sclerosing adenosis and proliferative fibrocystic disease [11, 16]. In these cases, the presence of tightly packed glandular structures, microcalcifications, apocrine metaplasia, and micropapillary projections can all contribute to histological confusion. If questions arise, comparison to the diagnostic biopsy or even knowing the morphologic subtype and grade of the prior tumor will help resolve the issue in the large majority of cases, although prior slides or reports are not always available.

When the histological picture is equivocal, as might be the case with cytologically bland well-formed glands, deferring to permanent section evaluation is appropriate. With paraffin-embedded tissue, immunohistochemical analysis may be performed to confirm the presence of myoepithelial cells (e.g., with antibodies to p63, smooth muscle myosin heavy chain) and/or basement membrane material (e.g., with antibodies to type IV collagen or laminin). The presence of myoepithelial cells and basement membrane material can reliably distinguish heterotopic epithelial inclusions from adenocarcinoma in most cases. Even then, however, exceptions do occur. Some squamous cysts lack myoepithelial cells, and a "revertant" form of metastatic ductal carcinoma in situ has been described, in which nests of metastatic tumor are actually surrounded by an intact basement membrane [11, 14, 17].

Most cases of local heterotopic epithelial inclusions involve collections of benign cells that appear to stably colonize lymph nodes. A few exceptions merit consideration, however. In these cases, ectopic cells are only present transiently in lymph nodes after trauma or location inflammation. For instance, there have been several published reports of both benign and malignant epithelium disrupted by trauma from surgery, biopsy, or even sutures or needles in women with breast cancer [18–20]. In one series, displaced benign or malignant tissue was found in lymph nodes from 12 of 43 consecutive women diagnosed with breast cancer by stereotactic core biopsy [19]. Fortunately, these fragments tend to be small and, in many cases, may only be recognized on permanent sections, when immunohistochemistry for cytokeratins can be employed to identify small clusters of ectopic epithelial cells. As such, they are unlikely to be a major pitfall during intraoperative evaluation of lymph nodes.

In the uncommon frozen section where these cell groups raise the possibility of metastatic disease, several histological features may help distinguish the displaced epithelium from true metastatic disease [20]. The clusters of displaced epithelial cells are characteristically small (i.e., usually less than 1 mm in diameter) and may consist of either benign or malignant epithelium. Like true metastases but unlike the stable epithelial inclusions considered above, they may be found in subcapsular sinuses. However, displaced clusters are distinguished from micrometastases by the presence of associated hemosiderin-laden macrophages and altered erythrocytes, which indicate recent trauma and are required for diagnosis as displaced epithelium. In practice, if one is uncertain at the time of frozen section analysis, discretion should be considered the better part of valor and deferral to a more thorough analysis on permanent sections should be considered.

Like heterotopic breast epithelia, hyperplastic mesothelium can be found ectopically in lymph nodes, especially mediastinal nodes [10, 21]. These cells are derived from tissues that normally line serosal surfaces in the abdomen and thorax. When present in lymph nodes, these benign cells can be mistaken for metastatic malignancy, including carcinoma or mesothelioma. The former is particularly important as lymph nodes are routinely removed prior to proceeding with potentially curative resection of the primary tumor in patients with non-small-cell lung cancer in order to clarify staging (see Chap. 5).

The incidence of hyperplastic mesothelial cells in lymph nodes appears to depend on clinical circumstance. Most reported cases have occurred in association with a serosal fluid collection [10], leading to speculation that normal mesothelium gains access to local lymphatics after being displaced into pleural, pericardial, or ascitic fluid. Consistent with this view, hyperplastic mesothelial cells were present in five of eight mediastinal lymphadenectomy specimens from patients with pericarditis or pleuritis but in 0 of 80 similar autopsy specimens [21, 22]. Given this association, patients with lung cancer accompanied by pleural effusions may be expected to be particularly at risk for false-positive lymph nodes due to hyperplastic mesothelium.

Grossly, lymph nodes containing hyperplastic mesothelium tend to be unremarkable [10]. However, several histological features are distinctive [10, 21]. Mesothelium is normally present as individual or small clusters of cells, although more impressive growth patterns with sheets of benign cells have been reported. The individual cells are polygonal or round, with centrally located, vesicular nuclei and small nucleoli. The cytoplasm has a characteristic biphasic appearance, with darker pink cytoplasm in the perinuclear region of the cell and lighter-staining eosinophilic cytoplasm more peripherally. Unlike most other benign mimics of metastatic cancer, hyperplastic mesothelium may show evidence of mitotic activity, and mitotic figures have been reported in as many as two per ten high power fields [10]. Like transiently displaced breast tissue, hyperplastic mesothelium is found primarily in the subcapsular sinuses, where metastases are also expected. In some cases, there may even be associated areas of necrosis, which can further complicate the interpretation [23]. Given these features, hyperplastic mesothelial cells are among the most difficult benign mimics to distinguish from metastatic carcinoma. In cases with mitotic activity and necrosis, comparison to the diagnostic biopsy may be the only reliable method for determining that the ectopic cells did not originate from the primary tumor.

The mechanism by which benign cells come to populate lymph nodes is not always clear. Two competing models have been proposed [11]. In the first model, ectopic cells are believed to originate in their normal, native tissues, before relocating to lymph nodes. This relocation is sometimes referred to as tissue embolization or so-called benign metastasis and is likened to the identification of viable trophoblast in pulmonary vessels in autopsies of women who died in the immediate postpartum period [24]. The second model supposes an embryonic origin for the ectopic cells in the lymph node. According to this model, inclusions arise from aberrant migration of embryonic precursors during early development. In the case of the transient displacement of breast tissue or mesothelium, the mechanism appears clearly embolic, as precipitating events or pathology are identifiable. The source of stable, local heterotopic epithelial inclusions is less clear. The preference for these inclusions to colonize the lymph node capsule complicates the benign metastasis model, given that malignant metastases have a predilection for growth in the subcapsular sinuses.

Müllerian Inclusions

Müllerian inclusions represent the second group of benign glandular inclusions found in lymph nodes. This condition represents a special case of endosalpingiosis in which Fallopian tube-like epithelium is found in lymph nodes. Despite the histologic relationship, Müllerian inclusions are believed to derive not from the uterine epithelium itself, but rather from peritoneal submesothelial stem cells, which may undergo metaplastic change in response to hormonal stimuli [25].

Müllerian inclusions are remarkably common in lymph nodes, occurring in 5–41% of autopsy and lymphadenectomy specimens from women and even rarely in men [26, 27]. Most commonly, they are found in subdiaphragmatic lymph nodes, which are generally only analyzed by frozen section in patients with prostatic or pancreatic adenocarcinoma. However, less commonly, Müllerian inclusions have also been identified in axillary nodes of women with breast cancer and misdiagnosed intraoperatively as metastatic breast cancer [9, 12, 28, 29]. The histological picture can be particularly confusing in cases of florid papillary endosalpingiosis, where the ectopic glands may have branching papillary structures and even mild atypia [29].

Again, careful microscopic evaluation of well-prepared frozen or permanent sections will typically reveal several features that distinguish Müllerian inclusions from metastatic disease [9, 28, 29]. Cystic glands are present and may be filled with proteinaceous secretions. In most cases, architecture will be distinct from the solid or complex patterns seen with adenocarcinoma. The lining of Müllerian inclusions is a monolayer of bland, cuboidal-to-columnar

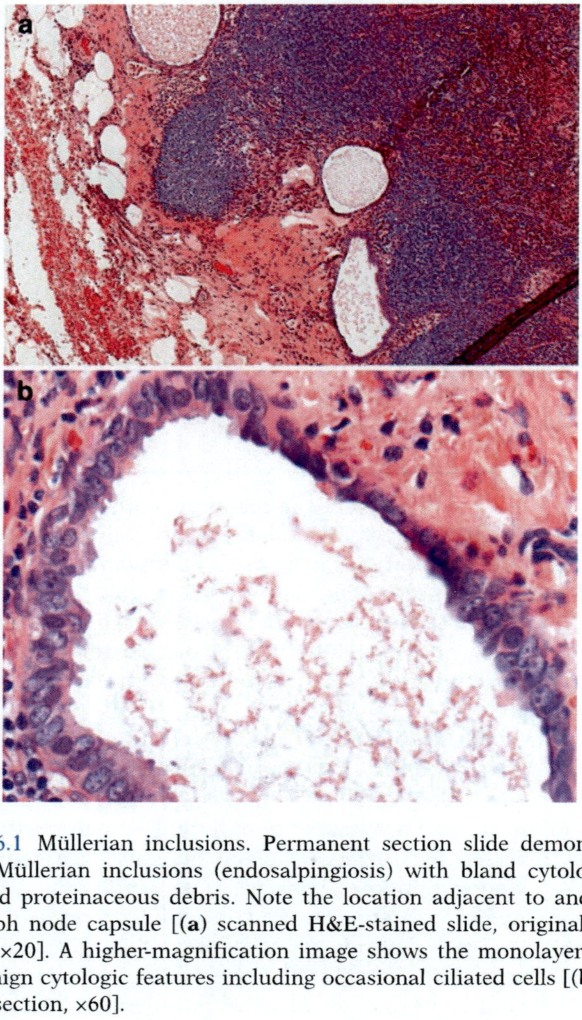

FIGURE 6.1 Müllerian inclusions. Permanent section slide demonstrating benign Müllerian inclusions (endosalpingiosis) with bland cytologic features and proteinaceous debris. Note the location adjacent to and within the lymph node capsule [(**a**) scanned H&E-stained slide, original magnification ×20]. A higher-magnification image shows the monolayer of cells with benign cytologic features including occasional ciliated cells [(**b**) H&E-stained section, ×60].

cells, which may or may not have cilia (Fig. 6.1). Mitotic figures are absent, and, in most cases, there is no atypia. As noted above, the exception to this rule is florid papillary endosalpingiosis. However, even then, atypia is mild by comparison to the more severe atypia likely to be observed in the primary tumor [29]. As a result, availability of the diagnostic biopsy for comparison can be invaluable. Finally, as with heterotopic epithelial inclusions, the location of the glands in question may be revealing. Whereas small foci of tumor cells are generally found in the subcapsular sinuses, Müllerian inclusions are more commonly identified in the fibrous capsule and perinodal fat.

In cases where a diagnosis must be deferred until permanent sections are available, immunohistochemical analysis can be useful. However, with conventional markers such as cytokeratins and hormone receptors, Müllerian inclusions will not always be distinguished from breast carcinoma [9, 29]. In these cases, markers such as mammaglobin or GCDFP-15 may be helpful [29].

BENIGN NONEPITHELIAL MIMICS OF NEOPLASIA

The cases above represent situations in which benign epithelial tissue can be confused with metastatic carcinoma in lymph nodes analyzed by frozen section. In other cases, the histologic picture may be complicated by nonepithelial cells in lymph nodes, which may have an epithelioid appearance that can be confused with metastatic tissue. Benign, nonepithelial mimics of carcinoma include nevus cell aggregates (NCAs), histiocytes and foreign-body giant cells, and megakaryocytes, as discussed below.

Nevus Cell Aggregates

In 1931, Stewart and Copeland first reported the presence of benign clusters of nevocytes in an otherwise normal lymph node [30]. Since that time, cases of what are now called "nevus cell aggregates" have appeared frequently in the literature. The incidence of NCAs is quite varied, depending on the clinical circumstance. Benign nevocyte inclusions are most common in patients with melanoma, with reported incidences ranging from less than 1 to 22% of lymphadenectomy specimens [31–33]. NCAs have also been reported in lymphadenectomy specimens from patients with other cancers, although at much lower frequency. For instance, Carson et al. identified just one example in 1,071 lymph nodes from axillary node dissections in women with breast cancer [33]. At the time of frozen section, these NCAs may be mistaken for metastatic carcinoma or melanoma. Because frozen section analysis of SLNs for melanoma is no longer recommended [34], it is in the SLN evaluation for breast carcinoma that NCAs are most likely to cause difficulty, despite the low frequency in that setting.

Several histological features distinguish NCAs from both metastatic carcinoma and melanoma [33, 35–39]. Benign nevocytes lack atypical cytologic features. They are uniformly small-to-medium sized, with a round-to-ovoid appearance. Likewise, nuclei tend to be round and centrally located, with inconspicuous nucleoli.

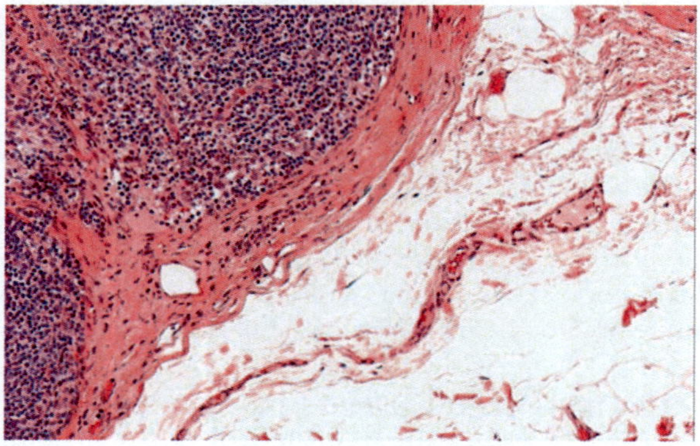

FIGURE 6.2 Nevus cell rest. Permanent section slide demonstrating a small nevus cell rest, with a group of epithelioid nevocytes situated in the fibrous lymph node capsule. Immunohistochemical studies confirmed the melanocytic nature of these cells (scanned H&E-stained slide, original magnification ×20).

Mitotic figures are rare. There is a small amount of homogenous, pale pink cytoplasm, and those melanosomes that are present may or may not be pigmented. Finally, cell borders are normally not distinct (Fig. 6.2). In terms of tissue architecture, NCAs may be arranged in nests or even cords, the latter of which may resemble the "Indian file" pattern classically associated with lobular carcinoma [36]. While this resemblance may create diagnostic confusion, location may clue the pathologist that the cells did not arise from the primary tumor. Whereas metastatic disease has a predilection for subcapsular sinuses, 93% of NCAs are present in the capsule. In the remaining cases, the ectopic nevocytes are almost always in lymph node trabeculae [33]. Because parenchymal involvement has been reported in rare instances [38], it may become necessary to defer to permanent sections, particularly if cellular features and comparison to the diagnostic biopsy do not yield a confident diagnosis. In cases where the diagnosis must be deferred, immunohistochemical studies on permanent section slides (if the small foci are still present) is usually revealing. While not informative for ruling out metastatic melanoma, S-100 and MART-1 can distinguish NCAs from metastatic carcinoma. Additionally, HMB-45 is usually negative or faint in nevocyte aggregates, and Ki-67 stains less than 1% of cells [37].

As with many mimics of metastatic carcinoma, the etiology of NCAs has not been clearly delineated. Paralleling theories of heterotopic epithelial inclusions, both embolic and embryonic origins, have been proposed. Von Albertini favored the former, arguing that NCAs represent benign metastases from cutaneous nevi in the catchment area of skin drained by the node in question [40]. Supporting this model, most NCAs occur in association with cutaneous nevi, which have been observed to project into the lumen of dermal lymphatics [41]. Moreover, multiple studies have reported an increased frequency of NCAs in SLNs of patients with melanoma. While this may reflect increased diagnostic attention to SLNs in some cases, it may also support a model in which growing neoplasms displace benign nevocytes into local lymphatics [33, 39, 42]. On the other hand, the tendency for NCAs to be found in the capsule or trabeculae rather than in subcapsular sinuses – an observation also seen with stable, heterotopic epithelial inclusions – potentially conflicts with an embolic/metastatic source. As a result, other investigators have favored a model in which lymph node nevocytes result from aberrant neural crest migration during embryonic development. However, it may be difficult to reconcile this model with the observation that NCAs have been found exclusively in superficial lymph nodes that normally drain cutaneous tissues [42].

Histiocytes and Foreign-Body Giant Cells

A variety of macrophage-lineage cells may also be mistaken for metastatic carcinoma in lymph nodes. Sinus histiocytes notoriously adopt an epithelioid appearance in subcapsular sinuses [14]. However, sinus histiocytosis is a normal and well-recognized reaction in the draining lymph nodes of cancers, and most experienced pathologists are adept at identifying conventional cases and separating them from intrasinusoidal infiltrates of carcinoma (Fig. 6.3). The picture becomes more complicated in circumstances where histiocytes adopt atypical or abnormal cytomorphologies, more easily confused with neoplastic cells.

Signet-ring histiocytosis represents a special case in which cellular features can prompt misdiagnosis [43–46]. This rare reactive disorder is seen in axillary and pelvic lymph nodes of patients with carcinoma. Interestingly, it has not been observed in lymph nodes from other locations, though the mechanistic significance of this observation is not clear. The characteristic cellular feature in this disorder is a distended vacuole that displaces the histiocyte nucleus, producing the signet morphology characteristic of some carcinomas and lymphomas [46]. Early reports suggested that this vacuole may be full of lipid material, because the first reported patient had diabetes mellitus [43]. However, there is little direct evidence to support this hypothesis.

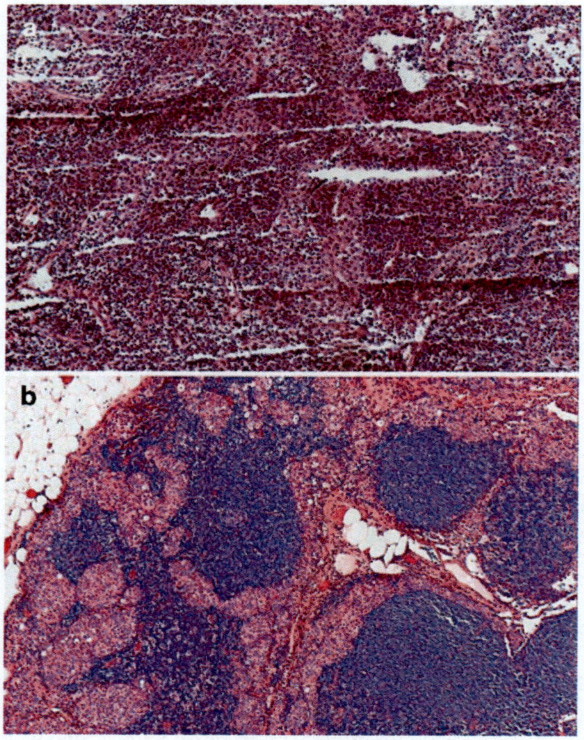

FIGURE 6.3 Sinus histiocytes and lobular carcinoma with sinusoidal pattern of metastasis. Actual frozen section of a lymph node demonstrating clusters of macrophages (histiocytes) within patent sinuses. Note the bland epithelioid morphology with small- to medium-sized nuclei and abundant cytoplasm [(**a**) scanned H&E-stained slide, original magnification ×20]. A permanent section slide of metastatic lobular carcinoma highlights a population of neoplastic cells with a sinusoidal distribution, which could potentially be confused with sinus histiocytes [(**b**) scanned H&E-stained slide, original magnification ×20].

Histologically, signet-ring histiocytes have a bland appearance, other than their characteristically displaced nucleus [46, 47]. There is no atypia, and nucleoli are either small or not discernible. Intracytoplasmic, eosinophilic inclusions may be present and may stain positively with PAS, further confounding the issue. Like metastatic adenocarcinoma and unlike many other mimics, signet-ring histiocytes are found in subcapsular and medullary sinuses.

Coupled with their abnormal cellular morphology, this localization can easily lead to misdiagnosis. Here, comparison to the diagnostic biopsy is particularly critical, although the diagnostic slides may not always be available. Although not helpful intraoperatively, immunohistochemical analysis will clarify the diagnosis. Signet-ring histiocytes are positive for CD68, CD45, lysozyme, cathepsin, and other histiocytic markers, but do not show immunoreactivity for cytokeratins or other markers of adenocarcinoma [46].

Another macrophage-derived mimic of metastatic carcinoma is granulomatous inflammation. This can be particularly concerning in the setting of frozen sections from mediastinoscopy procedures assessing for lung carcinoma. The presence of loose aggregates of epithelioid macrophages can mimic the presence of metastatic carcinoma, particularly metastatic squamous cell carcinoma. Careful attention to cytologic detail can help one avoid misdiagnosis of granulomata for metastatic carcinoma as the macrophages have bland nuclear features (Fig. 6.4a), indistinct cell borders, and lack mitotic activity. In contrast, metastatic squamous cell carcinoma has distinct cell borders, often with identifiable intercellular bridges, variable degrees of nuclear atypia including frequent prominent nucleoli, and mitotic activity (Fig. 6.4b). Necrotizing granulomata can also present a diagnostic dilemma, as metastatic carcinoma can also show extensive areas of necrosis with few viable-appearing malignant cells (Fig. 6.5). The difficulty in distinguishing metastatic carcinoma from granulomatous inflammation may be compounded in cases without a prior tissue diagnosis of carcinoma. Correlation with clinical history and prior slides (if available) may save one from falling into this relatively common potential pitfall.

Two other macrophage-lineage mimics of metastatic cancer merit brief consideration. Foreign-body giant cells have been noted to have an appearance that may be confused with micrometastases or isolated tumor cell groups in some cases (Fig. 6.6). The multinucleated giant cell bears a striking resemblance to a small focus of tumor cells with indistinct cellular borders. Complicating the picture, foreign-body giant cells can be found in the subcapsular sinuses, where micrometastatic disease is most likely [48]. While immunohistochemistry can be revealing at the time of permanent section, the only diagnostic pearl which can be helpful at the time of intraoperative consultation is the presence of polarizable material, which may or may not be identified in foreign-body giant cells but should be absent in true metastatic carcinoma.

The last macrophage-lineage mimic of metastatic disease to be discussed is the tattoo-pigment histiocyte. Although lymph nodes are not typically assessed intraoperatively in patients with melanoma

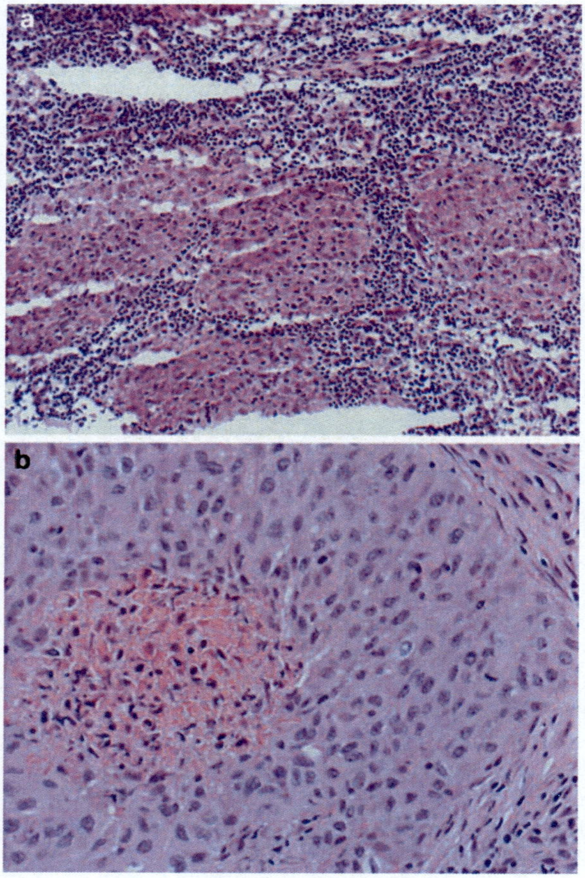

FIGURE 6.4 Nonnecrotizing granuloma and metastatic squamous cell carcinoma with necrosis. Actual frozen section slide with nonnecrotizing granulomata compatible with sarcoidosis. Note the low-grade cytology of the histiocytic population with evenly dispersed chromatin, inconspicuous nucleoli and abundant cytoplasm [(**a**) H&E, ×20]. This is compared with a permanent section slide prepared from the frozen section remnant that demonstrates a focus of metastatic squamous cell carcinoma with associated necrosis. Note the larger nuclei, irregular nuclear contours and vesicular chromatin, prominent nucleoli and scattered mitotic figures [(**b**) H&E, ×40].

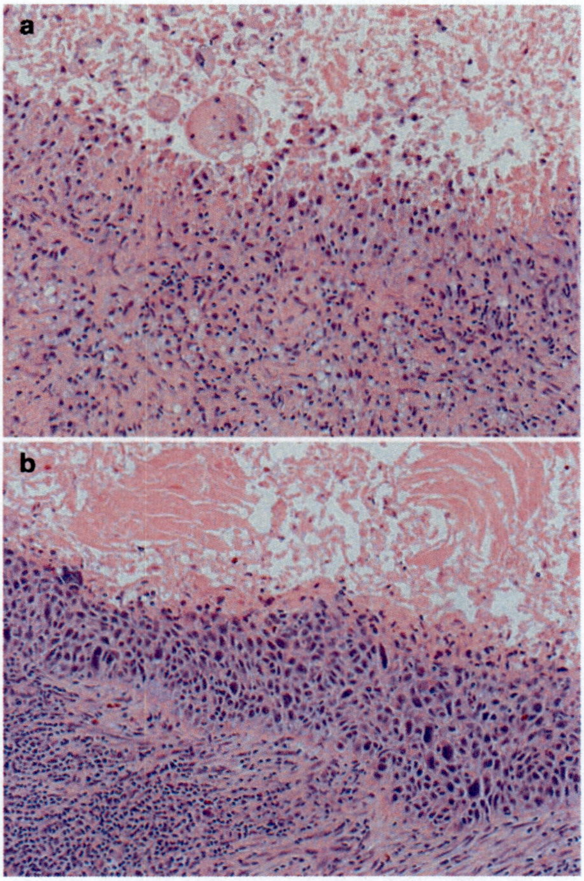

FIGURE 6.5 Necrotizing granulomatous inflammation and squamous cell carcinoma. Permanent section slide prepared from the frozen section remnant in a patient with squamous cell carcinoma of the lung. There is necrotizing granulomatous inflammation present with numerous histiocytes including multinucleated forms [(**a**) H&E, ×20]. In other areas of the lymph node, there is squamous cell carcinoma with associated necrosis that focally merged into areas of granulomatous inflammation. Note the high-grade cytologic features in the carcinoma cells [(**b**) H&E, ×20].

today, tattoo-pigment histiocytes previously represented another potential diagnostic pitfall [49–54]. Tattoo-pigment histiocytes can persist years after the original procedure and even after cutaneous markings have been removed by cosmetic procedures [52, 53].

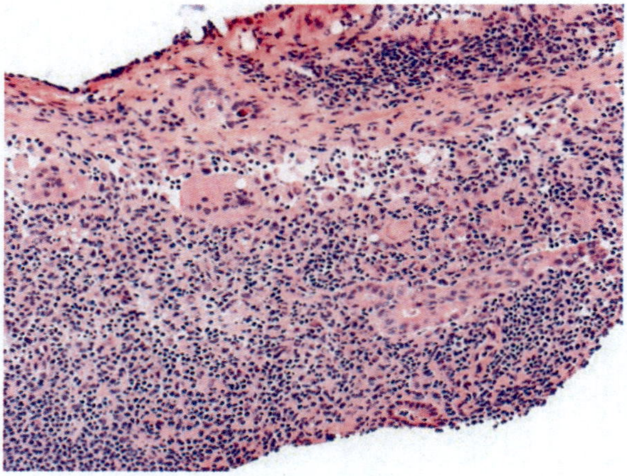

FIGURE 6.6 Micrometastasis with gland formation and associated multinucleated giant cell. Permanent section slide prepared from the frozen section remnant of a lymph node in which micrometastatic tumor was identified at the time of frozen section. Some morphologic features that distinguish the neoplastic glands in the capsule and lymph node parenchyma from the multinucleated giant cell in the subcapsular sinus include the arrangement of nuclei and orientation of the cells in the tumor focus and the disordered nuclei and friable, bubbly cytoplasm in the multinucleated giant cell (H&E, ×20).

Grossly, lymph nodes draining tattooed skin may be enlarged with a thin rim of dark pigment [52, 54]. Microscopically, the lymph node architecture is preserved, although there may be reactive follicular hyperplasia [53]. Black pigment is seen both inside of subcapsular macrophages and outside of cells within the sinuses. Additionally, the lymph node may show signs of fibrosis around the characteristic histiocytes, as part of a reactive effort to contain these cells (Fig. 6.7). Most characteristically and unlike metastatic melanoma, tattoo-pigment histiocytes are not mitotically active, which should allow an intraoperative distinction to be made. Nevertheless, clinicians should be encouraged to provide a thorough tattoo history whenever submitting lymphadenectomy specimens from melanoma patients.

Megakaryocytes

The final category of benign, nonglandular mimics of carcinoma metastases in lymph nodes occurs when megakaryocytes are misrecognized as neoplastic cells. Megakaryocytes may be present in

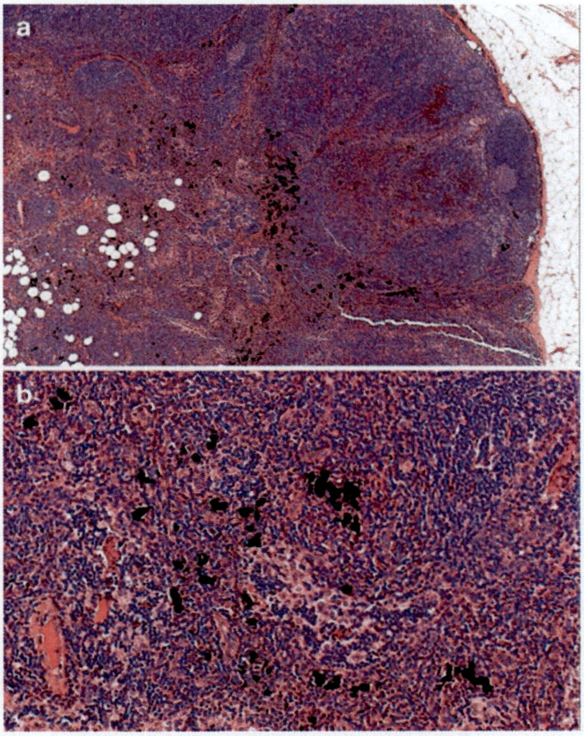

FIGURE 6.7 Tattoo pigment in macrophages may mimic metastatic melanoma. Permanent section slides of a lymph node demonstrating the presence of tattoo pigment in lymph node macrophages along the sinus trabeculae as well as in the lymph node parenchyma [(**a**) scanned H&E-stained slide, original magnification ×20]. Higher magnification shows that the pigment present in these cells obscures nuclear detail, precluding further assessment [(**b**) scanned H&E-stained slide, original magnification, ×20].

lymph nodes as an incidental finding or in cases of extramedullary hematopoiesis [55]. In the setting of extramedullary hematopoiesis, they are accompanied by other hematopoietic precursors, but these are not always readily appreciated. Histologically, megakaryocytes are identifiable as large cells, on an average of 20–25 mm in diameter and sometimes larger than 60 mm [56]. Nuclei are characteristically hyperchromatic and multilobulated, a picture which may be confused with severe atypia if the diagnostician is not considering the possibility of these hematopoietic cells. Megakaryocytes are usually found in lymph node sinuses, but unlike metastases, the favored locations are cortical or medullary rather than subcapsular [55]. The benign nature of these ectopic cells will become readily apparent on immunohistochemical analysis of permanent sections. Immunohistochemical studies for von Willebrand Factor, CD42b, CD61, and CD31 can help confirm the identity of megakaryocytes.

OTHER POTENTIAL PITFALLS

Dense Lymphoid Populations Mimicking Lymph Node Tissue

Confusion may sometimes arise when the tissue submitted to the frozen section room as "lymph node" is actually derived from some other tissue. In most cases, histological evaluation makes it obvious whether the tissue submitted is a lymph node or not. However, some benign tissues may resemble a lymph node involved by metastatic cancer. Two examples of this difficulty readily come to mind, both of which arise in frozen sections from the neck region. The first is distinguishing chronic lymphocytic thyroiditis from metastatic thyroid carcinoma in a lymph node and the second is distinguishing small fragments of thymic issue with their intrinsic squamous epithelial component from metastatic squamous cell carcinoma.

Lateral aberrant thyroid is heterotopic thyroid tissue found low in the neck and lateral to the jugular vein, separated from the body of thyroid [57]. Although once treated aggressively as metastatic disease, it is now appreciated that these tissues are often benign remnants of embryonic tissue that failed to fuse with the body of the thyroid [58]. In patients with chronic lymphocytic thyroiditis, lateral aberrant thyroid may also contain dense populations of lymphoid cells forming germinal centers and fibrosis [59]. These morphologic findings create a histologic picture that can be difficult to distinguish from metastatic thyroid carcinoma involving a lymph node, particularly if the tissue is labeled as a "lymph node" when submitted by the surgical team. Finding histologic characteristics of a lymph node, including a capsule with preserved subcapsular and trabecular sinuses, as well as scattered high endothelial venules, may help identify the tissue as a true lymph node, but the tissue may be small and fragmented, limiting histologic assessment.

Along a similar vein, thymic tissue may be misidentified as a lymph node involved by metastatic carcinoma, particularly when mediastinal nodes are assessed in patients with lung cancer [60] or in cervical lymph nodes from patients with carcinomas of the head and neck. The dense lymphoid population and associated epithelial elements make it possible to confuse thymic tissue for a lymph node

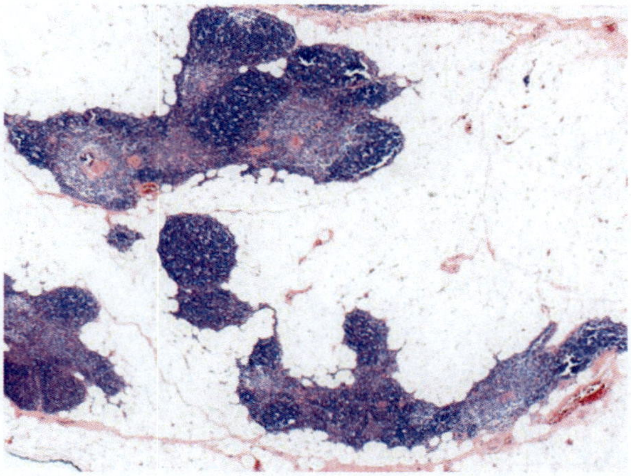

FIGURE 6.8 Normal adult thymus tissue. Permanent section slide of normal adult thymus tissue, showing the prominent lymphoid tissue that can occasionally be mistaken for lymph node tissue at frozen section. Note the lack of a fibrous capsule, sinuses or high endothelial venules, and the presence of extensive adipose tissue (H&E-stained section, ×10).

involved by metastatic disease. Lobules of thymic tissue, however, often have abundant admixed adipose tissue and although lobulated in appearance, lack a true capsule like lymph nodes (Fig. 6.8). In addition, the epithelial cell populations in benign thymic tissue are invariably benign cytologically without nuclear enlargement, hyperchromasia, or irregularities (Fig. 6.9).

Misleading or Incomplete History

An additional class of error results from poor communication between clinical teams and pathologists. Similar to the pathologist's responsibility to render a clear and accurate diagnosis, the clinical team has a responsibility when communicating relevant clinical history, as errors may arise when incomplete or inaccurate clinical information accompanies tissue submitted for evaluation. Frozen section assessment (and surgical pathology or cytology evaluations in general) is not to be treated as a game to see if the pathologist can render a diagnosis in a vacuum of clinical information [61]. Clinical context frames differential diagnoses, and good clinical histories help the pathologist direct diagnostic attention appropriately. When key information is lacking, differential diagnoses become unnecessarily narrow or broad, and interpretation suffers. It may

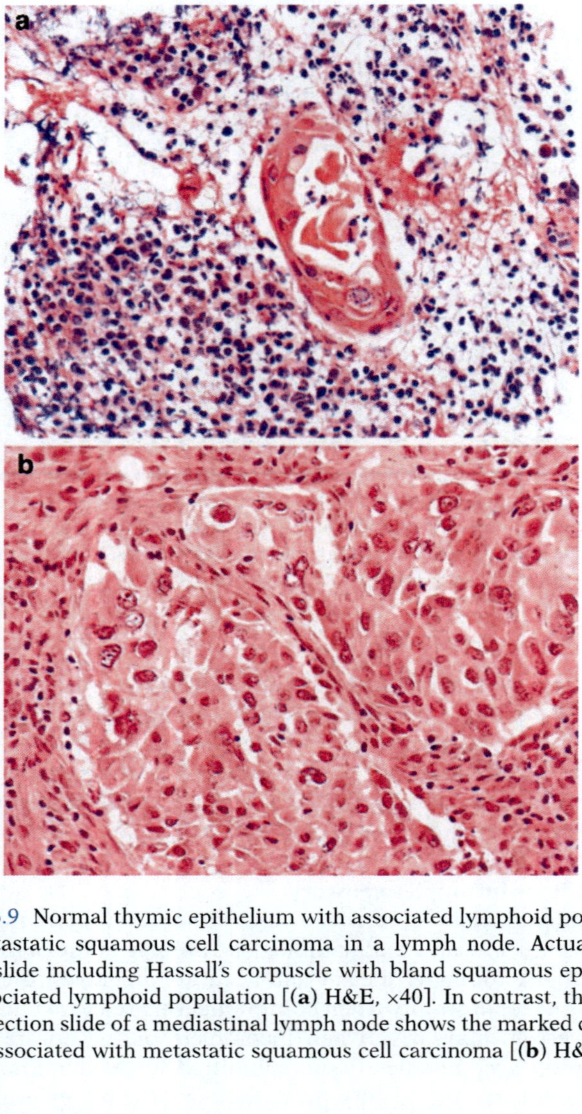

FIGURE 6.9 Normal thymic epithelium with associated lymphoid population and metastatic squamous cell carcinoma in a lymph node. Actual frozen section slide including Hassall's corpuscle with bland squamous epithelium and associated lymphoid population [(**a**) H&E, ×40]. In contrast, this actual frozen section slide of a mediastinal lymph node shows the marked cytologic atypia associated with metastatic squamous cell carcinoma [(**b**) H&E, ×40].

be unreasonable to expect the pathologist to accurately classify an abnormal cell population in the absence of appropriate clinical history. Particularly important is any prior history of malignancy, as the possibility of a new metastasis from a historically remote neoplasm must be considered in the differential diagnosis of a new metastatic process. Although the impetus for clarity, completeness, and accuracy of diagnosis ultimately lies with the pathologist, the clinical team must shoulder some of the burden by providing an accurate account of relevant history. Because one cannot always rely on this level of communication, a frozen section room setup with ready access to an online medical record can help avoid some of the shortcomings of histories provided on pathology requisitions.

Miscellaneous Causes of Error

Two additional causes of error may arise during interpretation of lymph node frozen sections and merit brief consideration. First, abnormal cell populations may be recognized but misclassified. As noted above, this error is particularly likely when appropriate clinical history is lacking. However, in other cases, it may simply reflect the experience and biases of the pathologist. Foucar and Foucar have written previously on sources of error in anatomic pathology [62]. While a comprehensive discussion of categories of error is beyond the scope of this chapter, some specific points made previously by these authors are worth emphasizing. While in some cases, the diagnosing pathologist may simply lack familiarity with the presenting entity (in other words, the pathologist may lack the knowledge necessary to make the diagnosis), in other cases one may steer away from the correct diagnosis as a result of any number of diagnostic biases. In particular, the clinical team's suspicions or the pathologist's personal interests and favored hypotheses may have undue influence on interpretation. The only remedy for such biases may be an appropriate awareness of one's tendencies and limitations so that one may guard against them.

Secondly and as discussed previously, high endothelial venules represent another potential source of diagnostic confusion. Because these structures consist of plump, cuboidal cells as opposed to the flat endothelial cells found in most venous structures, they may be confused with low-grade adenocarcinoma. Finding similar appearing structures in similar locations within the normal lymph node architecture, or finding lymphocytes or red blood cells within the lumen, may help confirm that the structures being evaluated are high endothelial venules. In contrast, if such tubules are associated with stromal desmoplasia, cytologic atypia, or mitotic activity, they are likely to represent metastatic disease and not normal high endothelial venules.

SUMMARY

The cases presented above represent a number of potential pitfalls that may be misdiagnosed as metastatic carcinoma in lymph nodes during intraoperative consultations. Frozen section specimens are notoriously of inferior quality relative to formalin-fixed, paraffin-embedded tissues, and in this setting, it can be difficult to distinguish convincing mimics from genuine metastases. However, the following histologic and cytologic features can be clues to help the pathologist to determine whether the ectopic cells are tumor derived or not. First, the benign mimics typically have no or, in rare cases, mild atypia. Comparison to the diagnostic biopsy can clarify this point and also allows a comparison of other cellular and architectural features. As a result, although it may be difficult logistically, it is recommended that appropriate slides be available in the frozen section room for comparison whenever possible. Second, mitotic figures are usually absent in the benign inclusions, although florid papillary endosalpingiosis represents one notable exception. Finally, location in the lymph node should always be considered. Early metastatic disease characteristically involves the subcapsular sinuses. Although a similar pattern is observed with transient heterotopic epithelial inclusions and sinus histiocytosis, many other mimics of metastatic disease are found disproportionately in the capsule, trabeculae, or in sinuses other than the subcapsular sinuses. When any of these features are appreciated, they should trigger caution and serve as a reminder to look for other histological signs that ectopic cell populations are only masquerading as metastatic disease.

REFERENCES

1. Voogd AC, Ververs JM, Vingerhoets AJ, Roumen RM, Coebergh JW, Crommelin MA. Lymphoedema and reduced shoulder function as indicators of quality of life after axillary lymph node dissection for invasive breast cancer. Br J Surg. 2003;90(1):76–81.
2. Flett MM, Going JJ, Stanton PD, Cooke TG. Sentinel node localization in patients with breast cancer. Br J Surg. 1998;85(7):991–3.
3. Van Diest PJ, Torrenga H, Borgstein PJ, et al. Reliability of intraoperative frozen section and imprint cytological investigation of sentinel lymph nodes in breast cancer. Histopathology. 1999;35(1):14–8.
4. Dixon JM, Mamman U, Thomas J. Accuracy of intraoperative frozen-section analysis of axillary nodes. Edinburgh breast unit team. Br J Surg. 1999;86(3):392–5.
5. Motomura K, Inaji H, Komoike Y, et al. Intraoperative sentinel lymph node examination by imprint cytology and frozen sectioning during breast surgery. Br J Surg. 2000;87(5):597–601.
6. Schwartz GF, Krill LS, Palazzo JP, Dasgupta A. Value of intraoperative examination of axillary sentinel nodes in carcinoma of the breast. J Am Coll Surg. 2008;207(5):758–62.
7. van de Vrande S, Meijer J, Rijnders A, Klinkenbijl JH. The value of intraoperative frozen section examination of sentinel lymph nodes in breast cancer. Eur J Surg Oncol. 2009;35(3):276–80.

8. Murayama H, Kikuchi M, Imai T, Yamamoto Y, Iwata Y. A case of heterotopic pancreas in lymph node. Virchows Arch A Pathol Anat Histol. 1978;377(2):175–9.
9. Piana S, Asioli S, Cavazza A. Benign mullerian inclusions coexisting with breast metastatic carcinoma in an axillary lymph node. Virchows Arch. 2005;446(4):467–9.
10. Argani P, Rosai J. Hyperplastic mesothelial cells in lymph nodes: report of six cases of a benign process that can stimulate metastatic involvement by mesothelioma or carcinoma. Hum Pathol. 1998;29(4):339–46.
11. Maiorano E, Mazzarol GM, Pruneri G, et al. Ectopic breast tissue as a possible cause of false-positive axillary sentinel lymph node biopsies. Am J Surg Pathol. 2003;27(4):513–8.
12. Peng Y, Ashfaq R, Ewing G, Leitch AM, Molberg KH. False-positive sentinel lymph nodes in breast cancer patients caused by benign glandular inclusions: report of three cases and review of the literature. Am J Clin Pathol. 2008;130(1):21–7. quiz 146.
13. Layfield LJ, Mooney E. Heterotopic epithelium in an intramammary lymph node. Breast J. 2000;6(1):63–7.
14. Resetkova E, Hoda SA, Clarke JL, Rosen PP. Benign heterotopic epithelial inclusions in axillary lymph nodes. histological and immunohistochemical patterns. Arch Pathol Lab Med. 2003;127(1):e25–7.
15. Colombo P, Cattaneo L. Axillary intranodal cysts associated with breast malignancy. Arch Pathol Lab Med. 2004;128(3):361–2.
16. Turner DR, Millis RR. Breast tissue inclusions in axillary lymph nodes. Histopathology. 1980;4(6):631–6.
17. Barsky SH, Doberneck SA, Sternlicht MD, Grossman DA, Love SM. 'Revertant' DCIS in human axillary breast carcinoma metastases. J Pathol. 1997;183(2):188–94.
18. Youngson BJ, Cranor M, Rosen PP. Epithelial displacement in surgical breast specimens following needling procedures. Am J Surg Pathol. 1994;18(9):896–903.
19. Youngson BJ, Liberman L, Rosen PP. Displacement of carcinomatous epithelium in surgical breast specimens following stereotaxic core biopsy. Am J Clin Pathol. 1995;103(5):598–602.
20. Carter BA, Jensen RA, Simpson JF, Page DL. Benign transport of breast epithelium into axillary lymph nodes after biopsy. Am J Clin Pathol. 2000;113(2):259–65.
21. Parkash V, Vidwans M, Carter D. Benign mesothelial cells in mediastinal lymph nodes. Am J Surg Pathol. 1999;23(10):1264–9.
22. Rutty GN, Lauder I. Mesothelial cell inclusions within mediastinal lymph nodes. Histopathology. 1994;25(5):483–7.
23. Rosai J, Dehner LP. Nodular mesothelial hyperplasia in hernia sacs: a benign reactive condition simulating a neoplastic process. Cancer. 1975;35(1):165–75.
24. Attwood HD, Park WW. Embolism to the lungs by trophoblast. J Obstet Gynaecol Br Commonw. 1961;68:611–7.
25. Lauchlan SC. The secondary mullerian system revisited. Int J Gynecol Pathol. 1994;13(1):73–9.

26. Farhi DC, Silverberg SG. Pseudometastases in female genital cancer. Pathol Annu. 1982;17(Pt 1):47–76.
27. Huntrakoon M. Benign glandular inclusions in the abdominal lymph nodes of a man. Hum Pathol. 1985;16(6):644–6.
28. Norton LE, Komenaka IK, Emerson RE, Murphy C, Badve S. Benign glandular inclusions a rare cause of a false positive sentinel node. J Surg Oncol. 2007;95(7):593–6.
29. Stolnicu S, Preda O, Kinga S, et al. Florid, papillary endosalpingiosis of the axillary lymph nodes. Breast J. 2011;17(3):268–72.
30. Stewart FW, Copeland MM. Neurogenic sarcoma. Am J Cancer. 1931;1031(11):301–7.
31. McCarthy SW, Palmer AA, Bale PM, Hirst E. Naevus cells in lymph nodes. Pathology. 1974;6(4):351–8.
32. Ridolfi RL, Rosen PP, Thaler H. Nevus cell aggregates associated with lymph nodes: estimated frequency and clinical significance. Cancer. 1977;39(1):164–71.
33. Carson KF, Wen DR, Li PX, et al. Nodal nevi and cutaneous melanomas. Am J Surg Pathol. 1996;20(7):834–40.
34. Scolyer RA, Murali R, McCarthy SW, Thompson JF. Pathologic examination of sentinel lymph nodes from melanoma patients. Semin Diagn Pathol. 2008;25(2):100–11.
35. Erlandson RA, Rosen PP. Electron microscopy of a nevus cell aggregate associated with an axillary lymph node. Cancer. 1982;49(2):269–72.
36. Fisher CJ, Hill S, Millis RR. Benign lymph node inclusions mimicking metastatic carcinoma. J Clin Pathol. 1994;47(3):245–7.
37. Lohmann CM, Iversen K, Jungbluth AA, Berwick M, Busam KJ. Expression of melanocyte differentiation antigens and ki-67 in nodal nevi and comparison of ki-67 expression with metastatic melanoma. Am J Surg Pathol. 2002;26(10):1351–7.
38. Biddle DA, Evans HL, Kemp BL, et al. Intraparenchymal nevus cell aggregates in lymph nodes: a possible diagnostic pitfall with malignant melanoma and carcinoma. Am J Surg Pathol. 2003;27(5):673–81.
39. Holt JB, Sangueza OP, Levine EA, et al. Nodal melanocytic nevi in sentinel lymph nodes. correlation with melanoma-associated cutaneous nevi. Am J Clin Pathol. 2004;121(1):58–63.
40. Von Albertini M. About a case of neuronaevus with lymph node metastasis of the same type. Bull Soc Fr Dermatol Syphiligr. 1935;2:1273–8.
41. Bell ME, Hill DP, Bhargava MK. Lymphatic invasion in pigmented nevi. Am J Clin Pathol. 1979;72(1):97–100.
42. Fontaine D, Parkhill W, Greer W, Walsh N. Nevus cells in lymph nodes: an association with congenital cutaneous nevi. Am J Dermatopathol. 2002;24(1):1–5.
43. Gould E, Perez J, Albores-Saavedra J, Legaspi A. Signet ring cell sinus histiocytosis. A previously unrecognized histologic condition mimicking metastatic adenocarcinoma in lymph nodes. Am J Clin Pathol. 1989;92(4):509–12.
44. Cappellari JO, Iskandar SS, Woodruff RD. Signet ring cell sinus histiocytosis. Am J Clin Pathol. 1990;94(6):800–1.

45. Frost AR, Shek YH, Lack EE. "Signet ring" sinus histiocytosis mimicking metastatic adenocarcinoma: report of two cases with immunohistochemical and ultrastructural study. Mod Pathol. 1992;5(5):497–500.
46. Guerrero-Medrano J, Delgado R, Albores-Saavedra J. Signet-ring sinus histiocytosis: a reactive disorder that mimics metastatic adenocarcinoma. Cancer. 1997;80(2):277–85.
47. Pathi R, Lawrence WD, Barroeta JE. Signet ring cell histiocytosis in axillary lymph nodes: a sheep in wolves' clothing? A potentially underrecognized pitfall in the diagnosis of metastatic breast cancer. Breast J. 2009;15(3):302–3.
48. Chiu A, Hoda RS, Hoda SA. Pseudomicrometastasis in sentinel lymph node-multinucleated macrophage mimicking micrometastasis. Breast J. 2001;7(6):440–1.
49. Anderson LL, Cardone JS, McCollough ML, Grabski WJ. Tattoo pigment mimicking metastatic malignant melanoma. Dermatol Surg. 1996;22(1):92–4.
50. Back L, Brown AS. Metastatic melanoma, or is it? Plast Reconstr Surg. 1986;77(1):138–40.
51. Moehrle M, Blaheta HJ, Ruck P. Tattoo pigment mimics positive sentinel lymph node in melanoma. Dermatology. 2001;203(4):342–4.
52. Chikkamuniyappa S, Sjuve-Scott R, Lancaster-Weiss K, Miller A, Yeh IT. Tattoo pigment in sentinel lymph nodes: a mimicker of metastatic malignant melanoma. Dermatol Online J. 2005;11(1):14.
53. Jack CM, Adwani A, Krishnan H. Tattoo pigment in an axillary lymph node simulating metastatic malignant melanoma. Int Semin Surg Oncol. 2005;2:28.
54. Peterson SL, Lee LA, Ozer K, Fitzpatrick JE. Tattoo pigment interpreted as lymph node metastasis in a case of subungual melanoma. Hand (N Y). 2008;3(3):282–5.
55. Hoda SA, Resetkova E, Yusuf Y, Cahan A, Rosen PP. Megakaryocytes mimicking metastatic breast carcinoma. Arch Pathol Lab Med. 2002;126(5):618–20.
56. Zafar N. Megakaryocytes in sentinel lymph node – a potential source for diagnostic error. Breast J. 2007;13(3):308–9.
57. Nishiyama RH. Overview of surgical pathology of the thyroid gland. World J Surg. 2000;24(8):898–906.
58. Kozol RA, Geelhoed GW, Flynn SD, Kinder B. Management of ectopic thyroid nodules. Surgery. 1993;114(6):1103–6. discussion 1106–7.
59. Acurio A, Taxy JB. Thyroid and parathyroid. In: Taxy J, Husain A, Montag A, editors. Biopsy interpretation: the frozen section. 1st ed. Philadelphia, PA: Wolters Kluwer; 2010. p. 171–92.
60. Gordon IO, Suzue K, Husain AN. Lung, mediastinum, and pleura. In: Taxy JB, Husain AN, Montag AG, editors. Biopsy interpretation: the frozen section. 1st ed. Philadelphia, PA: Wolters Kluwer; 2010. p. 47–79.
61. Ackerman LV, Ramirez GA. The indications for and limitations of frozen section diagnosis; a review of 1269 consecutive frozen section diagnoses. Br J Surg. 1959;46(198):336–50.
62. Foucar E, Foucar K. Medical error. In: Foucar K, editor. Bone marrow pathology. 2nd ed. Chicago, IL: ASCP Press; 2001. p. 76–86.

Index

C.H. Dunphy (ed.), *Frozen Section Library: Lymph Nodes*,
Frozen Section Library, 10, DOI 10.1007/978-1-4614-1253-3,